Digital Twin

The digital twin of a physical system is an adaptive computer analog which exists in the cloud and adapts to changes in the physical system dynamically. This book introduces the computing, mathematical, and engineering background to understand and develop the concept of the digital twin. It provides background in modeling/simulation, computing technology, sensor/actuators, and so forth, needed to develop the next generation of digital twins. Concepts on cloud computing, big data, IoT, wireless communications, high-performance computing, and blockchain are also discussed.

Features:
- Provides background material needed to understand digital twin technology
- Presents computational facet of digital twin
- Includes physics-based and surrogate model representations
- Addresses the problem of uncertainty in measurements and modeling
- Discusses practical case studies of implementation of digital twins, addressing additive manufacturing, server farms, predictive maintenance, and smart cities

This book is aimed at graduate students and researchers in Electrical, Mechanical, Computer, and Production Engineering.

Digital Twin

A Dynamic System and
Computing Perspective

Ranjan Ganguli
Sondipon Adhikari
Souvik Chakraborty
Mrittika Ganguli

CRC Press
Taylor & Francis Group
Boca Raton London New York

CRC Press is an imprint of the
Taylor & Francis Group, an **informa** business

First edition published 2023
by CRC Press
6000 Broken Sound Parkway NW, Suite 300, Boca Raton, FL 33487-2742

and by CRC Press
4 Park Square, Milton Park, Abingdon, Oxon, OX14 4RN

CRC Press is an imprint of Taylor & Francis Group, LLC

ISBN: 978-1-032-21362-0 (hbk)
ISBN: 978-1-032-21363-7 (pbk)
ISBN: 978-1-003-26804-8 (ebk)

DOI: 10.1201/9781003268048

Typeset in CMR10
by KnowledgeWorks Global Ltd.

Contents

Preface

Modeling and simulation have become the cornerstone of engineering and science for the last few decades. A large amount of research and development work has been directed toward computational methods for improving modeling. Such computer models are useful for system design and serve to mitigate the high costs of experiments and testing. However, there is a need to track the evolution of systems with time for diagnostics, prognostics, and life management. Degradation models of systems coupled with data emanating from sensors placed on the system allow the construction of digital twins, which permit tracking of a physical system in real time. The digital twin is an adaptive computer model of the physical twin which resides in the computer cloud.

In this book, we introduce digital twins using a physical twin of a spring-mass-damper system, a mathematical model for physical systems which is accessible to most engineers and scientists. Digital twin technology requires a comprehension of aspects from the fields of mechanical/aerospace engineering, electrical and communication engineering, and computer science. This book provides a background on these modeling and computer methods. The authors seek to present the material in a manner that a third-year university student of mechanical/aerospace engineering and a third-year university student of computer science/electrical engineering should both be able to read the book. This pedagogical approach ensures that the book is amenable to most engineers and scientists and to those in business and management but from a technology background.

The book introduces the computing and engineering background needed to actualize the concept of the digital twin. These include concepts such as sensors, actuators, the internet-of-things, cloud computing, estimation algorithms, high-performance computing, wireless communications, and blockchain, which have made the realization of the digital twin a possibility. Case studies from the literature are presented to elucidate the concepts. Further chapters provide material on dynamic system modeling, electrical analogs, probability and statistics, uncertainty modeling and quantification, and reliability and robustness of systems. A case study of a dynamic system is used to illustrate the concept of the digital twin. Surrogate models are then reviewed, and a Gaussian process approach is used to develop a surrogate model-based digital twin.

The book will be useful for senior undergraduates, graduate students, research scientists, and industry professionals seeking to understand the concept of digital twins. The book is also useful for researchers in engineering and science who want to contribute to the evolution of the field of digital twins.

Ranjan Ganguli
Sondipon Adhikari
Souvik Chakraborty
Mrittika Ganguli

Authors' Biographies

Dr. Ranjan Ganguli is currently a Senior Research Engineer at Viasat Inc. in Phoenix, USA. He received his MS and PhD degrees from the Department of Aerospace Engineering at the University of Maryland, College Park, USA in 1991 and 1994, respectively, and his B.Tech degree in Aerospace Engineering from the Indian Institute of Technology, Kharagpur, in 1989. He was a Professor in the Aerospace Engineering Department of the Indian Institute of Science, Bangalore from 2000-2021. He worked in Pratt and Whitney on engine diagnostics using machine learning during 1998-2000. He has completed sponsored research projects for companies such as Boeing, Pratt and Whitney, Honeywell, HAL, and others. His research is published in refereed journals and conferences. He has authored the books titled *Isospectral Vibrating Systems*, *Gas Turbine Diagnostics*, and *Engineering Optimization*, among others. He is a Fellow of the American Society of Mechanical Engineers, an Associate Fellow of the American Institute of Aeronautics and Astronautics, a Senior Member of the IEEE, and a Fellow of the Indian National Academy of Engineering. He also received the Alexander von Humboldt Fellowship and the Fulbright Fellowship in 2007 and 2011, respectively. He has held visiting scientist positions in Germany, France, and South Korea.

Prof. Sondipon Adhikari holds the position of Professor of Engineering Mechanics at the James Watt School of Engineering of the University of Glasgow. He received his PhD as a Jawaharlal Nehru Memorial Trust scholar at the Trinity College from the University of Cambridge. He was awarded the prestigious Wolfson Research Merit Award from the Royal Society (UK academy of sciences). He was an Engineering and Physical Science Research Council (EPSRC) Advanced Research Fellow and winner of the Philip Leverhulme Award in Engineering. He was the holder of the inaugural Chair of Aerospace Engineering at the College of Engineering of Swansea University. Before that, he was a lecturer at Bristol University and a Junior Research Fellow in Fitzwilliam College, Cambridge. He was a visiting Professor at the Ecole Centrale Lyon, Rice University, University of Paris, UT Austin, and IIT Kanpur, and a visiting scientist at the Los Alamos National Laboratory.

Professor Adhikari's research areas are multidisciplinary and include uncertainty quantification in dynamic systems, computational nano-mechanics, dynamics of complex systems, inverse problems for linear and non-linear dynamics, and vibration energy harvesting. He has published five books, more than 350 international journal papers, and 200 conference papers in these areas. Professor Adhikari is a Fellow of the Royal Aeronautical Society, an Associate Fellow of the American Institute of Aeronautics and Astronautics

(AIAA), and a member of the AIAA Non-Deterministic Approaches Technical Committee (NDA-TC). He is a member of the editorial board of several journals such as *Advances in Aircraft and Spacecraft Science, Probabilistic Engineering Mechanics, Computer and Structures*, and *Journal of Sound and Vibration*.

Dr. Souvik Chakraborty is currently working as an Assistant Professor at the Department of Applied Mechanics, IIT Delhi. He also holds a joint faculty position at the Yardi School of Artificial Intelligence, IIT Delhi. Dr. Chakraborty's research spans across a wide variety of topics including Scientific Machine Learning (SciML), stochastic mechanics, uncertainty quantification, reliability analysis, design under uncertainty, and Bayesian statistics, and he has published over 55 articles in peer-reviewed journals. Dr. Chakraborty received his PhD degree from Indian IIT Roorkee in 2017, and B.Tech and M.Tech degrees from the National Institute of Technology Durgapur and Bengal Engineering and Science University Shibpur (currently known as Indian Institute of Engineering Science and Technology Shibpur) in 2010 and 2013, respectively. Prior to joining IIT Delhi in 2020, he worked as a postdoctoral researcher at University of Notre Dame, USA and University of British Columbia, Canada from 2017 – 2019.

Mrittika Ganguli is a Principal Engineer and Director, Cloud Native Pathfinding in Intel's Network and Edge Architecture (NEX OCTO) team. She has 25+ years of experience in cloud hardware and software management, network and storage processing control and data plane, cloud orchestration, telemetry QOS, and scheduling Architecture. She is active in CNCF and Open Infra opensource initiatives and initiated a Service Mesh Performance (SMP) index called Meshmark. She has a MS in CS and 70+ patents and multiple IEEE papers in this area.

1 Introduction and Background

1.1 INTRODUCTION

A digital twin is an avatar of a real physical system which exists in the computer. In contrast to a computer model of a physical system which attempts to closely match the behavior of a physical system in a temporally static sense, the digital twin also tracks the temporal evolution of the physical system. The evolution of the computer replica with time is a key attribute of the digital twin. Some researchers have defined the digital twin at the conceptual level. However, these definitions have been very general due to the attempt to keep a very large number of systems within the ambit of the definitions. In this chapter, we seek to introduce the digital twin concept at a high level, while presenting the background needed to understand further chapters in the book.

A recent review of the literature on digital twins is provided by Tao et al [183]. Several definitions of the digital twin have been presented in the literature, two of which have become popular. Reifsnider and Mujumdar [158] consider digital twin to be a high-fidelity simulation integrated with an on-board health management system, maintenance history, and historical vehicle and aircraft fleet data. They wanted the digital twin to mirror the whole flying life of a specific operational physical twin. The availability of such a digital twin results in an enormous increase in reliability and safety for aircraft. Another popular definition of the digital twin was given by Glaessgen and Stargel [84]. They defined digital twin as an integrated multiscale, multiphysics probabilistic simulation of a complex product which uses the best available physical model, sensor updates, etc., to mirror the life of the physical twin. The first definition gives the view from the user perspective while the second definition provides a digital twin developer's perspective. The actual digital twin is an amalgamation of these two viewpoints.

The digital twin has theoretical foundations in information science, production engineering, data science and computer science [183]. Tao et al [183] identify four key aspects of digital twin research: (1) modeling and simulation, (2) data fusion, (3) interaction and collaboration and (4) service. While the focus in [183] was toward information science aspects, we classify these four aspects from the viewpoint of physical systems. The objective of the modeling and simulation phase of the digital twin is to create a virtual model which is a mirror image of the physical model [183]. For most physical systems, the virtual model will be a computer program which solves partial differential equations or matrix equations. This simulation model of the system must be

DOI: 10.1201/9781003268048-1

verified and validated, typically with experimental data. The virtual model may need to be updated at this stage and the model fidelity improved to minimize the discrepancy between the physical and virtual model. Uncertainty analysis of the virtual model is typically conducted to account for deviations in the physical system properties. Statistical measures can be used to quantify the deviation between the physical model and the virtual model, and optimization methods can be used to minimize this difference.

The second phase of the digital twin called data fusion involves collecting data from the system, typically using sensors. Some examples of sensors include pressure sensors, light sensors, accelerometers, gyroscopes, temperature sensors and motion sensors. In recent years, industrial sensors have become inexpensive and this has facilitated the development of the digital twin concept. The sensor data is then processed using signal processing, feature extraction, data mining, image processing and other methods, typically to amplify significant features of the data which reveal more about some desired condition of the physical system. Aspects related to the data having large size (big data) may be encountered at this stage because of a high rate of sampling by the sensors and the presence of a large number of sensors on the physical systems. Sensors can also be connected to prototyping boards such as the Arduino Uno and Raspberry Pi 2 which then allow them (sensors) to become part of the Internet of Things (IoT). Algorithms based on pattern recognition methods such as machine learning, fuzzy logic, etc., can also be used for data processing or feature extraction.

The third phase of the digital twin involving interaction and collaboration implies that there must be information flow between the physical model and the data fusion function of the digital twin. It is also possible that inputs generated by the virtual model are communicated to the physical model via actuators. Thus, the virtual model should incorporate changes in the physical model communicated through sensor data. The virtual model should be synchronized with the physical model at selected time steps (epochs) and must, therefore, evolve temporally along with the physical model. If possible, this synchronization should be done at frequent discrete time intervals and ideally in real time. Estimation methods such as Kalman filters or particle filters are useful during this stage to synchronize the physical and digital models. Since the virtual model typically exists as a computer program in the cloud, aspects of cloud computing become important along with the placement of the physical model in the context of the sensors as a part of the IoT [217].

The final function of the digital twin is service, which is the reason for which the digital twin exists, such as "structure monitoring, lifetime forecasting, in-time manufacturing etc." [183]. Estimation of service conditions and life of the physical system along with the prediction of required maintenance, downtime and replacement are some of the applications of the digital twin concept.

The discussion of the four phases of digital twin development shows that a plethora of technological domains are needed to develop an accurate digital twin. Several aspects of digital twin technology were addressed in earlier work on prognostics and health monitoring (PHM) [187, 191, 218], manufacturing [58, 89, 123] and other fields [225, 226]. However, the integration of these technology aspects is an onerous problem but many companies are extremely interested in the digital twin concept due to its ability to track expensive systems at low costs. Li et al [218] used a dynamic Bayesian network to monitor the operational state of aircraft wings. They replaced a deterministic model of the structure with a probabilistic model. Haag and Anderl [89] selected a bending beam test bench and created a digital twin for this system. The test bench consisted of a physical twin, a digital twin and a communication interface between the two.

Worden et al [210] suggest that digital twin is a powerful idea in the computational representation of structures. They mention that "there is currently no real consensus attempt to bring order to the subject of digital twin" and their paper seeks to make an attempt in this direction. To start with, they begin with a structure of the system which is the physical object which is placed in an environment. Both the structure and its environment are described by state vectors, and the digital twin is defined in terms of closeness between the structure and its twin or mirror image using an error bound. They point out that models for digital twin are often born as physics-based models. However, black-box models developed from data or gray box models developed from a combination of physics and data are also used to initiate the digital twin. As an example, one could start with a physics-based model and update the system parameters using system identification. While this paper presents several interesting ideas, it places them in a fairly abstract context. There is a need to define a digital twin for a simple system, such as a discrete dynamical system, which allows a concrete understanding of this concept. Such an example is also useful as a pedagogical tool which introduces the concept of a digital twin to engineering students and practicing engineers. We will address the digital twin of a discrete system in later chapters of the book.

1.2   MODELING AND SIMULATION

Modeling plays an important role in the development of a digital twin. Models can be broadly classified as real models, physics-based models and empirical models. A real or actual model of a system, for example, a business jet, would be a three-dimensional (3D) replica of the business jet in physical space. This replica could be a full-scale model or a scaled model which is used in wind tunnel tests. Sometimes, actual models are shown to illustrate a product before it has been earmarked for manufacturing. Physics-based models use knowledge about the system physics to create a mathematical model of the system. For example, a car could be modeled as a collection of springs, masses and dampers subjected to external forces. Physics-based models typically use equations to describe the system behavior. For example, the laws of mechanics should govern the behavior of a car and physics-based models may use Newton's laws of motion to develop equations for the system. We will study such physics-based models in detail in Chapter 3. Empirical models are developed from input-output relationships of a system. Sometimes, these empirical models are also called data-based models or identified models. For example, multi-dimensional curve fits using nonlinear functions seek to fit the function through data containing the input-output relations. These methods seek to use functions which interpolate well for data points not used to fit the curve. We will briefly review these types of models next and present an example for each.

Consider the motion of a swinging pendulum, a model which is often used in physics to illustrate simple harmonic motion. The pendulum is a good model for a swing and for several vibration mitigating devices. Pendulums are also sometimes used to keep time in clocks. The pendulum involves a mass m suspended by a string of length L. This system is then imparted an angle θ and then released. Using Newton's laws of physics, we can write the equation of motion of the pendulum as

$$mL\frac{d^2\theta}{dt^2} + mg\sin(\theta) = 0 \tag{1.1}$$

Here g is the acceleration due to gravity. We have assumed that there is no resistance due to air, the string is inextensible and the mass is sufficiently small such that it can be assumed to be a point mass or a particle, and there is no intrinsic damping in the system (for example, in the string) which would result in dissipation of energy in the system. Thus, Equation 1.1 represents a simple physics-based model of a pendulum. However, even this simple model is too complicated to be solved easily due to the presence of the term involving $\sin\theta$, which leads to a nonlinear ordinary differential equation. A further simplification can be obtained by assuming that the angle θ is small such that we can say that $\sin\theta \approx \theta$. Following this small angle assumption, we get a linear model

$$mL\frac{d^2\theta}{dt^2} + mg\theta = 0 \tag{1.2}$$

This differential equation can be solved to obtain

$$\theta(t) = \theta_0 \cos(\omega t) \tag{1.3}$$

where ω is the natural frequency of the system

$$\omega = \sqrt{\frac{g}{L}} \tag{1.4}$$

and T is the time period

$$T = \frac{2\pi}{\omega} = 2\pi\sqrt{\frac{L}{g}} \tag{1.5}$$

Equation 1.1 represents what could be called a high-fidelity physics-based model of the system relative to the low-fidelity physics-based model given by Equation 1.2. Physics-based models are useful for understanding a system but the results obtained may depart considerably from real-world situations due to the plethora of assumptions made for their development. Also, physics-based models which are finally solved in textbooks are considerably simplified to make them amenable to the mathematical tools available or known to most practitioners.

Data-driven models are often used in situations where physics-based models are not available and where input-output relationships exist between variables. Let us contemplate a situation where you recall from high school physics that the time period of oscillation of a pendulum depends on the length L of the string and the acceleration due to gravity g but do not remember the functional relationship. In that case, you could write a model

$$T = f_1(L, g) \tag{1.6}$$

This functional relationship would be a data-based model. You would need to perform experiments with different lengths L of the pendulum and in different gravity situations to generate input-output relationships needed to find the function f_1. Typically, some curve fitting methods could be used along with some chosen functions such as polynomials to find a good fit within the numerical range considered.

Now consider a case where you do not know the relation between variables for the simple physics-based model. You could surmise by observing the motion of pendulums that the time period would depend on the length L, mass m, gravity g and angle θ. This would lead us to a data-based model of the form,

$$T = f_2(m, \theta, L, g) \tag{1.7}$$

The function f_2 is unknown at this point and needs to be found to develop the data-based model. We can find this model using machine learning methods which can fit a nonlinear relationship between inputs and outputs. Chapter 3 will elaborate on such methods. The accuracy of the data-driven model largely depends on the functional form f_2 we select to fit the data, the variables selected for framing the model and the amount of data available in terms of data points linking the input variables to the output variables.

Consider a situation where we want to drop the assumptions of no aerodynamic drag and inextensible string. The possible inputs which could impact this system could be conjectured to be the mass of the pendulum m, the length of the string L, the acceleration due to gravity g, the angle θ, the density of air ρ, the viscosity of air ν, damping of the string c and stiffness of the string k. A data-driven model for the time period could have the form

$$T = f_3(m, \theta, L, g, \rho, \nu, c, k) \tag{1.8}$$

Here f_3 is a new and unknown function which needs to be found. Data-driven models have the advantage of allowing us to consider several parameters when developing the model and add new parameters easily. The data and the corresponding curve fit will reveal if some parameters are important or not. However, data-driven models are typically valid only in the vicinity of the numerical data used to develop the model. Data-driven models also have the reputation of being black boxes, especially when the functional representation becomes complicated as in neural networks. Physics-based models are more universal but often limited due to complex physics and mathematical difficulties. For example, we neglected many physical aspects to obtain a tractable physics-based model for the pendulum. Even the process of handling large θ leads to a solution in terms of special functions. Data-driven models are growing in importance as many internet-based systems have a plethora of data available from the users. However, data-driven models may need experiments to be performed by perturbing the variables across a range of values which can be difficult and time-consuming. Data-driven models become useful when the input-output relations are obtained as part of the operation and service of the system. Developing a data-driven model for the pendulum model would need experiments under different values of g, which is an onerous task [90]. Thus, physics-based models remain the cornerstone of engineering modeling, and measured data is typically used to improve the physics-based model.

Typically, physics-based models are parsimonious, or the simplest models available in a given situation, and often allow us to get a good understanding of the intricacies of the system. Parsimony is an important concept in science, which says that we should use the simplest available models and explanations as far as possible. In other words, the theory of parsimony says that the simplest possible explanation of a phenomenon or the simplest theory which fits the data should be selected. Unfortunately, ill-constructed and overly elaborate data-driven models may not be parsimonious though they may fit the data very well. This fact should be kept in mind when choosing models, be they physics based or data based. Overly complicated models may not interpolate well with new and noisy data.

Modeling is different from simulation. A simulation involves *using the model* to get outputs for a system from given inputs and therefore assists in creating a realization of the system, typically on a computer. Here, the model can be physics based or data based. The simulation uses the model to extract information about the system behavior through parametric studies and sensitivity analysis. A parametric study typically involves the process of varying one or two parameters of a system while keeping the others fixed so as to get a snapshot of the system behavior in a lower dimensional space more amenable to our understanding. Simulation also enables study of systems which are not easy to fabricate, for example, a helicopter in flight with a crack present in one rotor blade [78]. Such a simulation allows us to develop an idea about the system in a condition which would not be easy to replicate in reality. Simulations are widely used to plan space missions, and models play an enormous role in these situations. A reasonably good and validated model is a prerequisite for an accurate simulation. In many problems, uncertainties are present in the inputs. Treating the inputs as random variables and running the simulations at a plethora of points predicted by a statistical distribution is one way to check out the different scenarios which may befall a system. This procedure is known as Monte Carlo simulation and is widely used in system design and engineering to quantify the influence of randomness in the input variables on the system outputs. We will address stochastic simulation in Chapter 4 in the book.

Modeling and simulation allow us to predict the behavior of systems. However, the actual behavior of the system in the real world is typically different from the predicted value. Sensors measure data from the real system and this data can be used to validate the models. Validation is successful when the predicted measurements from a model match those measured by sensors closely. Model updating is the procedure by which model parameters are adjusted to match the measured output closely. Thus, sensors play a key role in the system design process.

1.3 SENSORS AND ACTUATORS

A sensor is a device which detects events or changes in the environment and sends this information to other electronics. The sensitivity of a sensor determines the change in the sensor output when the measured quantity changes. For example, a thermometer is a sensor which measures temperature. Suppose the mercury in the thermometer moves by 10 mm when the temperature changes by 1 degree Celsius. The sensitivity of the thermometer is 10 mm/C, assuming a linear variation. A sensor should have negligible effect on the environment it seeks to monitor. For example, if we insert a thermometer in a person's mouth, the reduction in temperature inside the mouth due to the thermometer should be insignificant. This can be often ensured by making the sensor as small as possible. Miniature sensors tend to be nonintrusive.

In addition to temperature, sensors can measure pressure, acceleration, displacement, humidity, air quality, smoke, light, alcohol, proximity and other important factors. Many such sensors are routinely found in homes, offices and vehicles. A sensor typically uses a physics-based change to measure the quantity of interest. For example, the mercury thermometer uses the fact that mercury expands or contracts due to changes in temperature. The process of mapping the change in the parameter (change in mercury level) to the actual measured output (the temperature reading) is called calibration. For example, the fact that water boils at 100 degrees Celsius can be used for calibration. An industrial temperature sensor is often called a thermistor. Most modern industrial sensors convert signals from some input to an electrical signal.

Sensors have been classified as active and passive. While active sensors need an excitation signal, passive sensors do not need an excitation signal and can directly and intrinsically generate the response. Sensors are also classified based on the domain of application or the technology used in the sensor. For example, we have biological, chemical, mechanical or electrical sensors. Another approach for sensor classification uses the energy conversion mechanism. For example, we have photoelectric, thermoelectric, electrochemical, electromagnetic and electromechanical sensors. We can also classify sensors as analog or digital sensors. Analog sensors work with continuous data while digital sensors work with discrete data. Typically, linear sensors are preferred for application since nonlinearity between input and outputs can create unwanted complexity. Most sensors are based on simple linear physical relationships.

The subject of sensors is quite vast and there is considerable literature available on this topic. Most of this work is based on developing sensors and addresses the underlying science which is used to convert the inputs to measurement signals which can be processed by the computers. Vetelino and Reghu [188] provide a good introduction to sensors. Sensors play a key role in the fructification of the digital twin concept. In particular, the advent of IoT has allowed sensors to be embedded into various machines and devices which are then able to communicate their current state in time to a central system. Simple threshold exceedance of sensor measurements are often used to increase the safety of systems. For example, a smoke sensor which is typically based on a measurement of airborne particles and gases can report an impending fire in a house, factory or establishment through the IoT. An oil debris monitor can find too many particles in the oil chamber in a machine and request amelioration. A review of some simple applications of sensors in daily life is given by Javaid et al [98].

In this book, we will consider mechanical systems to illustrate the concept of the digital twin. Mechanical systems undergo displacement and acceleration changes during their static and dynamic performance, respectively. Measuring acceleration and displacement is therefore important for mechanical systems. Accelerometers are sensors which convert mechanical motion into electrical signals. They are widely used to measure and control vibration and in the automobile and aerospace engineering fields. Gyroscope (gyro) sensors measure angular velocity and are widely used in navigation. Piezoelectric materials are often used in modern accelerometers and gyroscopes as they have excellent signal-to-noise capabilities. Piezoelectric sensors use the piezoelectric effect which converts motion resulting in strain in the material to an electric signal. These sensors are often available at miniature scales and come under the field called micro-electromechanical systems or MEMS [4].

In some situations, use of a sensor measurement can be made to predict another variable using an intermediate mathematical model. For example, strain could be measured and used to predict displacement and stress in a structure through the use of elasticity equations and material properties. Such a system could "measure" stress or displacement and can be classified as a virtual sensor [127] of stress or displacement. Signals emanating from sensors are typically fed to control systems or used to monitor the condition or health of the mechanical system. In the context of digital twin, sensor signals fed to the digital twin are used by estimation algorithm to calibrate and align the digital twin with the physical twin.

Sensor technology has greatly benefited from the advent of smart materials. Such materials typically convert one form of energy to another and can be used both for sensors and actuators. Typically, actuators are devices which perform a desired mechanical movement. For example, a piezoelectric actuator converts electrical signals to strain and thus displacement. Piezoelectric materials represent one of the most popular smart materials. Magnetostrictive materials convert magnetic fields to displacements and vice versa. Shape memory alloys convert temperature changes to displacement. Electrorheological (ER) fluids can be used to control the damping of a system through the application of an electric field. Magnetorheological (MR) fluids can be used to control the system damping through a magnetic field. A background on smart materials can be found in Bengisu and Ferrara [17].

While many actuators use smart materials with electric field-based input, others use hydraulic or pneumatic approaches. Electric actuators are typically simple, have low cost and noise, and are energy efficient. However, electric actuators may be unsuitable for heavy loads. In contrast, hydraulic actuators are good at handling high loads but often have a number of parts such as valves, tanks, pumps, regulators and pipes. More parts make a device amenable to problems. A pneumatic actuator allows for rapid movement between any two points. However, they need valves, tubes and a compressor. The appropriate actuator selected depends on the problem and the domain of application.

Electric actuators convert electricity to kinetic energy, for instance, through an electric motor. Hydraulic actuators typically have a fluid motor which uses hydraulic power to create mechanical motion. Since liquids are hard to compress, hydraulic actuators can generate large forces. Most hydraulic actuators use a piston to convert the fluid pressure to mechanical force. Pneumatic actuators typically convert energy in compressed gas into mechanical motion. The gas (often air) exerts pressure on a piston to generate mechanical motion.

Actuators can also be classified as linear or rotary depending on the time of motion they produce. Linear actuators move in a linear plane, while rotary actuators produce rotary motion in a circular plane.

Digital twins rely on sensors to sense or monitor the state of the system. In some cases, the digital twin may be required to initiate action to automatically amend the condition of the system. Actuators allow the system to initiate corrective action to mitigate any problems. Some of these functions are already performed by modern control systems which take information from sensors, process this information on a computer and then send signals to actuators to take corrective action. The digital twin addresses a particular system (such as an aircraft with a given tail number) and uses its life cycle historical evolution to tailor the actuation actions specifically for the aircraft. This customization of the control sytem greatly enhances its power.

1.4 SIGNAL PROCESSING

Modern systems are equipped with a large number of sensors which routinely collect measurements. However, these measurements typically contain noise and outliers. The measurements are also obtained at discrete intervals of time which is sometimes called as epoch or discrete time and indicated by an integer k. When a time series of measurements $x(k)$ is obtained over $k = 1, 2, 3, \ldots, N$, we say that a signal has been generated. Removing noise and outliers from this signal while ensuring that the key features of the signal are kept intact is one of the main goals of the field known as signal processing. Sometimes, this problem is also referred to as data processing or data smoothing. Many algorithms have been developed for noise removal, denoising or filtering of the signal. We will introduce some simple filters which are useful for removing noise from signals.

A finite impulse response (FIR) filter can be represented as [76]

$$y(k) = \Sigma_{i=1}^{N} b(i)x(k - i + 1) \tag{1.9}$$

where $x(k)$ is the kth measurement, $y(k)$ is the kth output, N is the filter length and $b(i)$ are the weighting coefficients. The weighting coefficients define the behavior of the filter and sum to one. Various FIR filter designs are possible by tailoring the weights of the filter. A simple moving average filter can be obtained by considering that all weights are equal. This mean or average filter is widely used in data smoothing. For instance, a 10-point mean filter has the form

$$y(k) = \frac{1}{10}(x(k) + x(k - 1) + x(k - 2) + \ldots + x(k - 9)) \tag{1.10}$$

Here each of the ten weights are equal to $1/10$. The moving average is good for removing Gaussian noise from a signal. Gaussian noise has a probability distribution function identical to that of the normal distribution. It is a good model for the noise present in many physical systems.

Another simple filter is the exponentially weighted moving average (EWMA) which is a popular IIR (Infinite Impulse Response) filter. This filter can be expressed as

$$y(k) = ax(k) + (1 - a)y(k - 1) \tag{1.11}$$

The parameter a is a smoothing parameter whose value ranges between 0 and 1. Values of a ranging from 0.15 to 0.25 are widely used in applications. The exponential filter has memory as it retains the entire time history of the signal through the use of $y(k - 1)$.

Both the FIR and IIR filters are linear filters which can be used to remove Gaussian noise from signals. However, sometimes outliers which are of non-Gaussian origin are also present in signals. An outlier is a point which differs considerably from the other data in a given data set. Median filters are an effective tool for removing non-Gaussian outliers. A median filter with a length or window of length $N = 2K + 1$ is defined as

$$y(k) = median(x(k-K), x(k-K+1), \ldots, x(k), \ldots, x(k+K-1), x(k+K)) \tag{1.12}$$

The median filter typically operates on an odd number of samples. Consider a 5-point median filter which has $K = 2$. This filter can be written as

$$y(k) = median(x(k-2), x(k-1), x(k), x(k+1), x(k+2)) \tag{1.13}$$

The median filter needs past and future values of the measurements to give an output. In practical terms, the filter has a time lag.

A simple data smoothing algorithm can be developed by first smoothing the signal using the nonlinear median filter to remove outliers and then using the linear FIR or IIR filter to remove Gaussian noise. However, we should be careful of not removing important features present in the signal by over smoothing. There is a tradeoff between noise removal and feature preservation. Many advanced signal processing methods are available in the literature and there are a plethora of books on this topic. A good introduction is provided by Orfanidis [140].

The sensors when coupled with IoT systems create a key technology for digital twins. Accuracy of the sensor measurements play an important role in ensuring that the digital twin closely follows the real physical system. Thus, signal processing can play an important role in assuring that the signals received by the information processing algorithms have high signal/noise ratios and can be relied upon to yield good estimation results.

1.5 ESTIMATION ALGORITHMS

Estimation is the process of finding an approximation or estimate of a quantity from input data which is uncertain, incomplete or of doubtful integrity. In many problems, sensor data is noisy and the system model may have approximations or errors. However, we need to estimate the value of output quantities in these nonideal settings. Estimation methods play an important role in performing this process. A widely used estimation method is the Kalman filter.

The Kalman filter estimates states of a linear dynamical system perturbed by Gaussian noise. It is an optimal estimator. We will illustrate the Kalman filter as used in a gas turbine diagnostic problem [76]. Consider a measurement vector z and a state vector x. The state vector represents the parameters which define the current situation or "state" of the system. For example in gas turbine diagnostics, we could have four measurements and five states. The measurements could be exhaust gas temperature, low rotor speed, high rotor speed and fuel flow. These measurements are found in most aircraft engines. The states could be flow capacities of the fan, low pressure compressor and high pressure compressor, and efficiencies of the low pressure and high pressure turbines. Typically, a gas turbine has five modules named the fan, low pressure compressor, high pressure compressor, high pressure turbine and low pressure turbine. A linear model linking the measurements and states could be obtained in terms of the influence coefficient matrix H.

$$z = Hx + \nu \tag{1.14}$$

Consider that x and ν are Gaussian and independent. The optimal estimation problem then involves minimizing

$$J = \frac{1}{2}(z - Hx)^T R^{-1}(z - Hx) + (x - \mu_x)^T P^{-1}(x - \mu_x) \tag{1.15}$$

The first term in the right hand side of this equation measures the measurement error, weighted by the inverse of the noise covariance R^{-1} and the second term measures the state error weighted by the inverse of the state covariance P^{-1}. The optimal estimator is then given by

$$\hat{x} = [P_0^{-1} + H^T R^{-1} H]^{-1}(H^T R^{-1} z + P_0^{-1}\mu_x) \tag{1.16}$$

Putting the optimal estimator in predictor-corrector form yields

$$\hat{x} = \mu_x + [P_0^{-1} + H^T R^{-1} H]^{-1} H^T R^{-1}[z - H\mu_x] \tag{1.17}$$

which can be written as

$$\hat{x} = \mu_x + P_0 H^T [H P_0 H^T + R]^{-1}[z - H\mu_x] \tag{1.18}$$

The first term in the right hand side (μ_x) is the predictor and the second term is the corrector. The term $[z - H\mu_x]$ is called the residual. The gain is given by $P_0 H^T [H P_0 H^T + R]^{-1}$ is called the gain. Writing in a generalized form,

$$\hat{x} = \bar{x} + D(z - H\bar{x}) \tag{1.19}$$

The discrete time Kalman filter can now be defined as

$$x_k = \phi(k)x_{k-1} + \omega_k \tag{1.20}$$

where x_k is the state vector at epoch k, ϕ_k is the state transition matrix and ω_k is the process noise vector. Alongside the system model, a measurement model is also present

$$z_k = H_k x_k + \nu_k \tag{1.21}$$

Here z_k is the measurement vector, H_k is the geometry matrix and ν_k is the measurement noise vector. Several assumptions have been made in this derivation.

1. The noise vectors ν_k and ω_k are Gaussian and zero mean
2. $R_k = cov(\nu_k, \nu_k) > 0$
3. $Q_k = cov(\omega_k, \omega_k) \geq 0$
4. $Cov(\omega_k, \nu_j) = 0$ i.e., there is no correlation between process and measurement noise
5. $\mu_x = E(x_0)$, which means that the initial guess of the state is known
6. $P_0 = cov(x_0, x_0) = P_0 > 0$

The discrete time Kalman filter is then given by

$$\hat{x}(k+1|k) = \phi(k+1)\hat{x}_k \tag{1.22}$$

$$P(k+1|k) = \phi(k+1)P_k\phi^T(k+1) + Q_{k+1} \tag{1.23}$$

$$D_{k+1} = P(k+1|k)H_{k+1}^T P(k+1|k)H_{k+1}^T + R_{k+1})^{-1} \tag{1.24}$$

$$\hat{x}_{k+1} = \hat{x}(k+1|k) + D_{k+1}(z_{k+1} - H_{k+1}\hat{x}(k+1|k)) \tag{1.25}$$

$$P_{k+1} = [1 - D_{k+1}H_{k+1}]P(k+1|k) \tag{1.26}$$

Let us consider what these five equations mean.

1. The first equation extrapolates the state vector x from the kth epoch to the $k+1$th epoch. The transition matrix is the operator which facilitates this extrapolation.
2. The second equation extrapolates the covariance matrix P from the kth epoch to the $k+1$th epoch. The transition matrix and process noise facilitate this extrapolation.
3. The third equation involves calculation of the Kalman gain.
4. The fourth equation is the state update using the Kalman gain.
5. The fifth equation updates the covariance.

The Kalman filter is a powerful estimator and had an elegant mathematical form. However, it is a predictor-corrector method and needs a good starting guess to find a good estimate. In that sense, the Kalman filter is similar to the Newton-Raphson method widely used for the solution of nonlinear algebraic equations. A key aspect of the Kalman filter is the calculation of the numerics which are contained in the P, Q and R matrices. The matrix H about the system or process itself is also crucial. In gas turbine diagnostics, the H matrix is obtained from influence coefficients which come from a linearized model of the engine at a given operating state. Thus we see that a baseline model, typically physics based, is needed by the Kalman filter.

Kalman filters are robust to measurement and process noise and can also account for missing and faulty sensor measurements. Details about Kalman filters can be obtained from books such as the one by Zarshan and Mushoff [222] and a simpler one more amenable to beginners by Kim [105]. In the context of digital twins, Kalman filters can play a useful role in estimating the state of the phyical twin through noisy measurements. However, a growing trend uses machine learning for this problem and we will focus on this modern approach in the current book.

We should point out the existence of other estimators such as the particle filter and the unscented Kalman filter. As we have discussed before, the Kalman filter is optimal for a linear system with Gaussian noise. When a system is nonlinear, the Kalman filter can still be used for state estimation. However, in these settings, the particle filter may give better results albeit at the cost of higher computational expense. On the other hand, if a system has non-Gaussian noise, the Kalman filter is the optimal linear filter. However, the particle filter may perform better. The unscented Kalman filter can provide a middle ground between the computational efficiency of the Kalman filter and the superior performance of the particle filter. Details about the particle filter and the unscented Kalman filter can be obtained from Simon's book on optimal state estimation [174].

1.6   INDUSTRY 4.0

Digital twin literature is often replete with the use of the term "Industry 4.0" which we seek to explain in this section in terms of evolution of the technological society. The history of the industrial age is classified in terms of industrial revolutions. The first industrial revolution involved the change from the use of hand production approaches to the use of machines. The use of steam power is often used as a typical example of the first industrial revolution. The second industrial revolution emanated from the creation of transportation networks such as railways and also the use of telegraphs which permitted rapid communication of information. The third industrial revolution was created by the development of the computer and is also known as the digital revolution. Machines started to take over many functions performed by humans during this stage. Industry 4.0 is an idea for creating a fourth industrial revolution by the synthesis of computers and manufacturing. This merger of computers, internet, humans and machines has also led to the growth of what are sometimes called cyber-physical systems, a key component of Industry 4.0. Four design principles have been identified as critical for Industry 4.0 [141]. These can be enumerated as

1. Interconnection—This concept involves the ability of machines, sensors, electronics and people to connect and communicate with one another, typically through the internet. The IoT plays a key role in this technology.
2. Information transparency—A huge amount of data is created through the use of IoT and sensors at different stages of the manufacturing process across the life cycle of a product. This data could be used to improve the manufacturing process and even design.
3. Technical assistance—The systems are able to help humans with decision making, problem solving and to releive people from performing onerous or unsafe tasks.
4. Decentralized decisions—The cyber physical systems are imbued with autonomy and can make decision is a decentralized manner with minimum

assistance needed from humans. The human assistance is only sought in extreme situations.

While Industry 4.0 is more of a technology vision, there are large elements of this vision which need digital twin technology to fructify. For example, mobile devices, human-machine interfaces, smart sensors, data analytics, augmented reality and the IoT are key components of the Industry 4.0 vision. There is a need here of a virtual copy of the physical system to be developed. Therefore, we can say that digital twin technology is a prerequisite for the successful implementation of the Industry 4.0 vision. However, while digital twin technology is all-encompassing system level technology, Industry 4.0 is more focused on manufacturing and the factory infrastructure.

1.7 APPLICATIONS

Digitals twins need to be applied in real-world applications to become useful. There have been several cases where digital twin technology has been applied across different industries. We will give an overview of some of these applications as mentioned in recent literature.

1.7.1 MAINTENANCE

Digital twins are attractive for maintenance applications and represent one of the main areas of their application. Numerous applications of digital twin methods are available in the literature. Typically, such applications can be classified into several parts [68]

1. Reactive maintenance typically involves an emergency such as that caused by a breakage or breakdown. This maintenance strategy is only used in situations where the system breakdown will not have a large impact on business or life. Otherwise, there are deleterious consequences in terms of system downtime, need to replace the system or loss of life.
2. Preventive maintenance is often followed to mitigate the problems caused by reactive maintenance. A proactive approach is used to perform maintenance activities at periodic intervals such that service interruptions can be avoided. Preventive maintenance is better than reactive maintenance for systems which are important and/or have high costs. For example, an automobile which is checked and serviced every six months has a much less likelihood of sudden failure. However, over-maintaining the assets can cause higher costs as each periodic and often mandatory service entails expenditure. In many cases, the preventive maintenance activities may result in revenue for the servicing firms which may therefore be satisfied with this approach.
3. Condition-based maintenance involves anticipating maintenance activity using information about degradation or deviation of the system from its normal or baseline state. Typically, condition-based maintenance is related

to diagnosis and seeks to find anomalies in the system by using thresholds on sensor measurements and Artificial Intelligence (AI)-based approaches to novelty detection. A key feature of condition-based maintenance involves sensor placement and basic data cleaning and threshold monitoring software.

4. Predictive maintenance involves collecting information about the system in terms of usage and then using physics-based or phenomenological models to predict the remaining life. Phenomenological models are typically developed using experimental data and are often used to model degradation in materials such as composites which have complex physics. Predictive maintenance is also called prognosis and involves prediction of the time point where service is needed and where the product becomes unsafe to operate. One objective of predictive maintenance is to reduce the costs which can come from preventive maintenance. This allows the time between mandated maintenance to be reduced which can reduce the operating costs of the system. As an example, newer automobiles have various sensors indicating when oil, tyres, etc., need to be replaced and have reduced number of routine maintenance checks. A key aspect of predictive maintenance is model development and models can be analytical, physical or numerical. These models often estimate how the system degrades with time and accounts for different environmental conditions experienced by the system as obtained from the sensor suite. It has been found that predictive maintenance is one of the most important topics being investigated in the Industry 4.0 vision.

5. Prescriptive maintenance seeks to optimize maintenance based on predictions. This approach seeks to build on predictive maintenance by presenting an action plan for the system or product being monitored. Prescriptive maintenance benefits most from the use of digital twin technology as availability of sensor data in real time can be used to develop updatable models of the system which can then create an action plan suitable for the specific system under consideration. For example, an automobile can use the internet to schedule an appointment with the shop when its prescriptive maintenance system is indicating the need and would also inform the user about the same. It may also inform the user about the need to change their driving habits, fuel preferences and best routes to take for a journey by using the model, sensor and information processing sub-systems as well as information of gas stations and service stations along the route which it gets from the internet.

Digital twins are typically not used in the context of reactive maintenance strategies since this simplistic approach focuses on repairing the system once it has broken down. Here, digital twins can be used for a postmortem analysis of the reasons behind the system failure. If a digital twin is available for such a system, the maintenance approach should quickly be changed to predictive or prescriptive to avoid the massive disruptions caused by the reactive approach.

Typically, preventive maintenance plans are developed with the help of experts familiar with the system or provided by the system manufacturer. In many instances, over-maintenance is carried out due to ideas of enhancing system safety or productivity, a procedure which has even been called as "abusive preventive maintenance" [68]. A validated digital twin could be used to chart out a better preventive maintenance plan. However, validation of the digital twin is important as use of an inaccurate and unvalidated digital twin can lead to deleterious consequences.

Once sensors are incorporated into the system, as in the case with condition-based monitoring, updating the digital twin becomes possible and predictive maintenance can be used. Here the predictive models can be integrated into the computer simulation underlying the digital twin. Based on the incoming sensor data, the digital twin can use estimation or machine learning to predict the degradation of a component or the system. However, predictive models often need historical data for calibration and validation. Data on system improvement after the repair event must be fed back to the digital twin to keep it aligned with the physical twin.

Digital twin is necessary for prescriptive maintenance. Here, the data from the sensors constantly feeds into the digital twin and updates the predictive and optimization models inside the digital twin to keep it aligned to the physical twin. The digital twin in turn is able to predict impending degradation and failure of the system and to suggest palliative action.

The move from reactive maintenance to prescriptive maintenance requires a steep improvement in monitoring capability through sensors, modeling capability of the system and the degradation mechanisms, estimation and information processing needed to update the models based on the sensor data and finally predictive capability development, calibration and validation. Also, many preventive maintenance practices are mandated by the regulatory bodies and need to be followed irrespective of the presence of digital twin capability. The digital twin technology should be brought within the gambit of the regulation and certification processes.

We now consider some specific application ideas for digital twins to aircraft structural life prediction. Tuegel and his co-workers [187], in a seminal paper, proposed the digital twin concept as applied to an aircraft with a specific tail number 25-0001. This important idea adds specificity and shows how the digital twin differs from modeling which is typically applicable to all specimen of a given aircraft. This physical twin 25-0001 has a digital twin 25-0001D/I which starts as a high-fidelity finite element/computational fluid dynamic model of the aircraft. The digital twin should be made as realistic as possible in terms of geometric and manufacturing detail and should include any anomalies specific to this particular physical twin 25-0001. The digital twin accepts probabilistic inputs of "loads, environmental, and usage factors". The digital twin undergoes usage damage using damage models. A second digital twin model called 25-0001D/A is then created with is linked to the sensor system on the aircraft. For example accelerations, temperatures and pressure readings from the aircraft flights are recorded and then fed back to the structural model for the 25-0001D/A. The system then uses Bayesian statistical methods to update 25-0001D/I periodically to reflect actual usage of the aircraft 25-0001D/I. We see that this idea imagined as a concept some time ago (2011) is still quite useful. Nowadays, we can consider 25-0001D/I as the baseline physics or data-based model which is born alongside the physical twin but in the virtual world. Due to the internet and the cloud allowing continuous sensor measurements, data processing, estimation and feedback, this baseline model morphs into the digital twin 25-0001D/A as time passes. The digital twin starts alongside the physical twin and evolves with it.

The technologies identified by Tuegel [187] for the development of digital twin for aircraft life prediction are

1. Multiphysics modeling
2. Multiphysics damage modeling
3. Integration of system model and damage model
4. Uncertainty quantification, modeling and control
5. Manipulation of large, shared databases
6. High resolution structural analysis capability

We see that these technology challenges remain critical to the use of digital twin concepts for aircraft structural maintenance. We shall address some of these areas later in the book.

1.7.2 MANUFACTURING

There is enormous potential for the application of digital twin in smart manufacturing. Three examples can be considered as given by Shao and Helu [172].

1. A "machine health twin" can be developed to monitor the condition of manufacturing equipment and predict faults and prospective failures in the equipment. It is possible that the machine may take action to ameliorate its condition based on the data obtained and the subsequent data analysis.
2. A "scheduling and routing twin" can be developed to gather data from shop floor systems such as production equipment and enterprise resource planning (ERP) systems. This data is then analyzed to evaluate the current state of the production system and predict fluctuations in inventory, customer demand and resources. Cycle time reduction, resource optimization and inventory cost reduction could then be performed with the help of this digital twin.
3. A "commissioning twin" can use data collected by monitoring new equipment performance during commissioning to facilitate system optimization.

A recent review paper on the application of digital twin technology to manufacturing found several key application themes [50] by carefully studying the literature in the area. The digital twins are used to

1. Support the production system management
2. Monitor and improve the production process
3. Support the lifecycle of a machine process
4. Handle the flexibility of the production system
5. Perform maintenance
6. Help for safety reasons in human-robot interactions
7. Design a machine
8. Evaluate the performance of a cloud-based digital twin

We see that many aspects of manufacturing can be improved by using digital twin technology. A concrete example of digital twin in manufacturing was developed by Haag and Anderl [89]. They created a bending beam test bench to demonstrate the digital twin concept. The physical twin consists a bending beam clamped between two linear actuators. They also integrated two load cells into the holding fixture on one side to measure the resulting force. The displacement was calculated as the difference between the position of the two linear actuators. Thus the physical system is a widely understood mechanical system for which the material and geometric properties and sensor measurements are available. This allows for simple development of the digital twin using modeling. A baseline digital twin was developed using a computer aided design (CAD) model of a bending beam. The physical twin and digital twin were then connected using a messaging protocol which allowed IoT connectivity. Thus, through the internet, the physical twin communicated the measured displacement and force values to the digital twin which updated itself using a finite element simulation. This test bench, through simple in concept, illustrated the concept on digital twin and could be extended to more complex systems such as an aircraft wing.

1.7.3 SMART CITIES

Digital twin concept has been applied for the management of smart cities [164]. A key feature of this technology is the use of 3D visual models. While earlier work addresses visualization of the builidings, the new work focuses on using 3D models to obtain information about the urban landscape. For example, potential solar panel deployment for optimal use of renewable energy can be indicated by the city digital twin. A semantic 3D city model decomposes the objects into parts based on logical and not graphical considerations. By studying the city of Helsinki in Finland, the authors [164] were able to provide open energy data to all stakeholders. The city model was developed using CityGML where GML is a geography markup language. Some information provided by the developed system was

1. The solar power potential and suitable locations for solar panels

2. The heat loss taking place from the roofs
3. The potential of green roofs (existing and for the future)

These goals clearly tie into the larger goals for alleviating climate change and improving energy efficiencies of cities, which is central to sustainability.

The data from the past was obtained from data catalog datasets. The data from the present is obtained from IoT sensors. The future of the system is predicted by the digital twin. The authors [164] mention that "in the smart city concept, the digital twin can be a specific block or district". The digital twin allows users to test scenarios such as the impact of speed limits imposed on vehicles in a location on noise levels or air quality. The smart city digital twin has a plethora of data coming from numerous sensors at high rates and the relationships between input and output are not known a priori. Therefore, machine learning can be useful for developing the digital twin of smart cities.

Software plays a key role in the deployment of digital twin across complex systems. For example, data stored in MS Excel or in pdf files which is available to only a single city department can be amalgamated into a digital twin. Software such as Amazon Web Services (AWS) plays an important role in facilitating the process of developing the digital twin in the cloud. Such software systems include capabilities such as storage, compute, AI and machine learning, IoT, etc. Much of digital twin implementations for real systems are guided by software such as AWS, Microsoft Azure, etc. We will review these pieces of software and other computing aspects in the next chapter. Later chapters will largely be devoted to the mathematical modeling and machine learning aspects of the digital twin as illustrated by a dynamic system. However, a knowledge of the computer aspects is necessary to fructify the digital twin technology in the real world.

2 Computing and Digital Twin

The idea of using modeling and simulation for the analysis, design and optimization of engineering systems is now ubiquitous. However, these ideas were applied for systems in general and not for a specific specimen of a system, such as a particular aircraft. Digital twins have become possible due to the internet of things (IoT), wireless communications, high-performance computing capability in the cloud framework and the ability to handle big data using signal processing and machine learning. Furthermore, areas such as cybersecurity have become important for the successful and safe deployment of digital twins. In this chapter, we will briefly cover these technologies and provide references for further study. Some of the material in this chapter is adapted from References [194–201].

Digital twin is defined as a virtual representation of a physical system. Digital twin can be easily represented in software as a simple multi-dimensional matrix array. In many instances, these arrays having multiple dimensions are called tensors. While the creation of the digital twin often needs the solutions of physics-based models such as partial differential equation, the computer considers the digital twin to be a tensor representation of the physical reality. This chapter addresses the aspects of the digital twin which are closely related to computer implementation. In fact, this compute content is critical for the successful manifestation of digital twin in the industry. Later chapters will focus on the mathematical infrastructure behind the digital twin concept.

While the application of the digital twin concept ranges from an electro-mechanical sub-system to a smart city to a drone to a human body, there is a need for multiple level software and hardware frameworks, orchestrated systematically to assist in twin model building, real-time analysis and updates based on physical twin real-time changes. Figure 2.1 shows a simple framework for developing a digital twin. As shown, telemetry collection from sensors on the physical system requires an IoT defined framework including a system of databases and data-lakes, analysis using machine learning and deep learning which can be hosted at the end-point or in the cloud on popular cloud provider solution sites such as Amazon AWS, Microsoft Azure and Google GCE.

2.1 DIGITAL TWIN USE CASES AND THE INTERNET OF THINGS (IoT)

Digital twins are used for electrical systems, aeronautical systems such as gas turbine engines, drones, helicopters, space systems such as satellites, a Mars

DOI: 10.1201/9781003268048-2

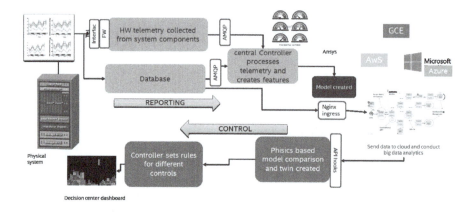

Figure 2.1 A digital twin generation framework.

rover, and automotive systems such as a car chassis, semiconductor chips used inside the car, buildings, bridges, dams, power plants, nuclear reactors, the human body and many more. Such systems generate a continuous stream of data from IoT [204] sensors that measure information about environmental conditions and performance metrics. Internet of things (IoT) is defined as a system of sensor-control center-actuator system (Figure 2.2). While earlier only computers were connected to the internet, mobile devices expanded this space dramatically. The advent of 4G, 5G and satellite technology allows systems ranging from washing machines, air conditioners to cars, aircraft engines and airplanes to be connected to the internet, thereby allowing easy flow of data between the system and the data processing computer infrastructure.

The raw and unprocessed data collected from the sensors are converted to streams of digital data and transferred over the wireless network. The analog signals emanating from the sensor are converted to digital numeric values by a digital acquisition system. The wireless network for IoT sensors' digitized data transfer follows a specified set of protocols. The data is forwarded through

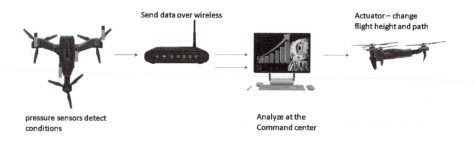

Figure 2.2 An example IoT system for drones.

multiple nodes, and using a gateway, the data is connected to other networks such as wireless Ethernet.

IETF RFCs for routing such data through a gateway of devices have been developed [203]. While the network could be prone to information loss, it should still be able to transport enough data to make conclusions. The wireless network between thousands of IoT devices and the edge servers connects all the sensors and is extremely sophisticated. To alleviate over-subscription delays and congestion when large amount of data is streamed out and also to mitigate the risk of system crashes when there are a limited number of active connections, an asynchronous architecture is recommended. This eliminates the requirement of having a single thread per device and handles the control events sent by each one, forwarding them back to the simulation.

2.2 EDGE COMPUTING

The data collected from the IoT sensors and streamed out through the wireless networks needs to be stored and analyzed to form a digital twin. As mentioned before, the amount of this data is large due to the number of sensors, the sampling rate of the data and the large number of devices being monitored. Moving such an enormous amount of data to cloud is not always practical and certainly onerous. Furthermore, there is a lag in model creation and updates and slow reaction times. Therefore, the twin generation framework can be implemented at a smaller scale with enterprise-grade infrastructure and modern IT concepts, close to where data is captured such as a central location within the city premises for a smart city twin or the manufacturing floor for a factory twin or product twin. The hardware and software framework which enables capturing of data and processing and analyzing near the source of data is known as Edge Computing.

Edge computing infrastructure is described as a "mesh network of micro data centers that process or store critical data locally and push all received data to a central data center or cloud storage repository, in a footprint of less than 100 square feet", according to research firm IDC. You can differentiate the edge into 3 types.

1. The IoT or Near Edge
2. Operational Edge (like manufacturing)
3. Far or Enterprise edge (Telco)

An enterprise-grade edge computing strategy establishes the foundation for digital twins. Edge Computing allows organizations to mitigate the substantial costs, network bandwidth constraints and latencies associated with transferring enormous amounts of data to and from a cloud data center for processing and analysis.

IoT devices making use of edge computation have become a critical component of building realistic digital twins. An edge infrastructure not only supports data collection, but also facilitates industrial IoT applications for predictive maintenance, preventing operational downtime, developing innovative products and improving the customer experience. The edge network data centers consist of servers that are scattered more frequently across the world compared to the cloud data centers which are routinely placed in remote locations for the lower cost. IoT Devices local to the edge servers connect and send and receive data to the edge server without having to communicate with cloud servers. This makes the data communication process more efficient.

To manage surges in demand effectively, as data is streamed in, edge networks require a distributed load balancing system. A typical microservices and service mesh environment typically ensures that the network always has sufficient compute to manage large influxes of data and simultaneously process this data without crashing. In addition, the load balancing system must address the logging, analysis and debugging of processes within the network. A network visualizer, for example, can provide enormous amounts of information about the connection between micro services. A typical service mesh environment will also help in managing latency, bandwidth, and throughput with detailed statistics, such as the lost packets, the window sizes or the time since the last send or receive reflecting the asynchronous patterns of data being collected. A typical software solution which handles ingest of data, processing in parallel with security and logging would be microservices with proxy architecture as outlined in [200]. The step-by-step process will include:

1. Creating virtual visualizations of physical assets using historical and real-time data. Filtering data and reducing the cost of data traffic, which becomes more important as the number of sensors increase to help generate a more precise virtual object. A multi-access edge computing—(MEC) - could be used if the digital twin traffic is over a wide area network.
2. Formulating a digital twin model using the principles and analysis methods as outlined in the rest of the chapters. These chapters are focused on differential equation-based modeling of the dynamic system and machine learning-based surrogate modeling.
3. Generating simulations and employing real-time analysis, plus interacting more with the physical twin in an automated way.
4. Twin to twin interactions using AI and machine learning.

Edge is useful here also if real-time analytics and use of machine learning, can no longer be performed on local devices due to meager processing power nor in the cloud due to latency issues. This need for edge computing grows as the digital twins start to communicate with one another and require low levels of latency to ensure processes are not jeopardized. This is also likely to be on-premise edge (e.g., in factory) or MEC. Figure 2.3 shows a good mapping of where the solution should be based on time on one hand versus performance requirements on the other hand as a key requirement.

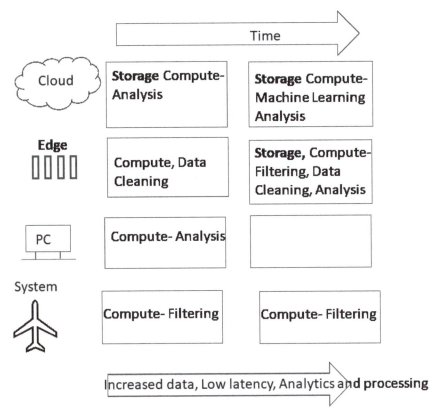

Figure 2.3 Solution choice of computer system based on time and performance requirements.

2.3 TELECOM AND 5G

A telecom network's role is crucial to the data collection, creation and streaming of audio, video and text data part of the digital twin model. Note that all these types of data need to be converted to multi-dimensional arrays of numbers or tensors before being transmitted. Digital twins, edge computing and AI will continue to accelerate the need for more connectivity as they generate more traffic. Private LTE and/or 5G highlight the challenge with the telecommunication business model. One reason companies are amenable to the idea of a private network is that it means they do not need to pay per GB per sensor—a model that is not feasible in a digital twin world. Moreover, these manufacturers cannot put their mission-critical and highly sensitive data on the shared, public telecom network. Data security is an important area in digital twin technology.

Let us consider the benefits of 5G and 3GPP standards for digital twin creation. Most importantly, the requirements of a sophisticated network that we alluded to earlier in this section is what 5G ultra wide-band (UWB) and MEC provides, namely:

1. Low latency: 200 milliseconds for 4G, 1 millisecond(1ms) with 5G and 5g speed of 20 Gigabits-per-second (Gbps) peak data rates and 100+ Megabits-per-second (Mbps) average data rates
2. Millimeter wave and band frequency spectrum: Certain vendors have 700-1000 MHz band frequencies and 20–30 hz mm-wave frequency
3. Multi-access edge computing (MEC): This is a network architecture concept which enables cloud computing capabilities and an IT service environment at the edge of any network. Mobile edge computing provides both an IT service environment and cloud-computing capabilities at the mobile edge of the network, within the radio access network (RAN) and near mobile subscribers, enterprises and other organizations

Central software architecture: Figure 2.4 shows an adoption of a software framework which assists on data collection and reduction at host, twin creation and data storage in the cloud.

As evident from Figure 2.4, the authors of [196] represent data collected by physical sensors at the lowest level and argue the need to quickly form a model of that. That virtual model will have its own data store in the cloud or edge. The relationship between the physical entities and virtual ones are maintained and used to transform any twin model and visualize in the intelligent server layer.

The proposed architecture adheres to the SenAS [202] model, where the data are generated by the "things" and are finally consumed by the humans or by other machines. All the data that are useful to improve the Quality

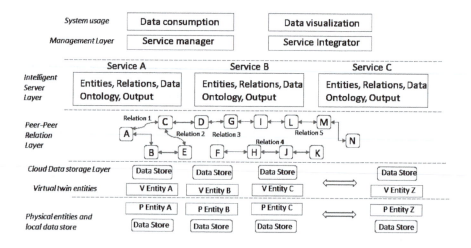

Figure 2.4 A framework for digital twin creation and visualization.

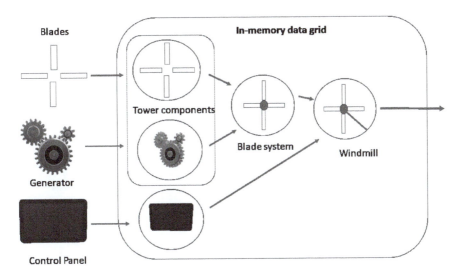

Figure 2.5 Digital twin data represented in in-memory data store.

of Service (QoS) of the physical things are stored in the cloud-based Data Center. An in-memory data grid (IMDG) is a suitable data structure that resides entirely in RAM (random access memory), and is distributed among multiple servers. Data collected in SenAS model can be transformed into an IMDG structure and a digital twin hierarchy created as shown in Figure 2.5 which shows a hypothetical windmill.

2.4 CLOUD

The lowest-level digital twins now are hosted at the edge next to their corresponding devices without any changes to the code. The higher-level digital twins continue to run in the cloud or on-premises—wherever the required computing resources are located. For hosting and building a digital twin at the cloud there are two powerful cloud solutions presented by Microsoft Azure and Amazon AWS IoT. These two approaches are discussed next.

2.4.1 MICROSOFT AZURE

Azure Digital Twins provides a platform as a service (PaaS) offering that enables the creation of twin graphs based on digital models of entire environments (Figure 2.6). While using Azure a solution would include

1. Creating a model to represent the physical entities (as in Figure 2.4) of the physical environment to create a custom twin type called model [197] using a JSON-like language called Digital Twins Definition Language (DTDL).

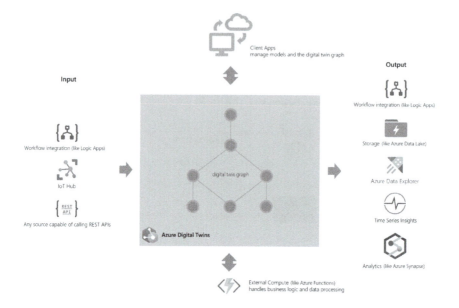

Figure 2.6   Azure reference architecture for digital twin.

2. Creating a live execution environment or a live graph using the relationships in the custom DTDL model, connect twins representing the environment and visualize the Azure Digital Twins graph in Azure Digital Twins Explorer. Using a rich event system, the graph can be kept current with data processing and business logic which arrives from the next steps 3 and 4.
3. Connecting the model to IoT and IoT Edge device using the IoT Hub. These hub-managed devices are represented as part of the twin graph and provide the data that drives your model.
4. Analyzing the data collected using Time Series Insights, storage, and analytics. Time series history of each twin can be tracked, and the Time Series Model can be aligned with the source of the twin in Azure [195].

An example of a basic model, written as a DTDL interface [198], is given next. This model describes a windmill, with one property for an ID. The windmill model also defines a relationship to a farm model, which can be used to indicate that a farm twin is connected to certain windmill twins.

Telemetry describes the data emitted by any digital twin, whether the data is a regular stream of sensor readings or a computed stream of data, such as occupancy, or an occasional error or information message. A sample data point such as temperature can be defined as:

JSON

(
"@id": "dtmi:com:adt:dtsample:home;1",
"@type": "Interface",
"@context": "dtmi:dtdl:context;2",
"displayName": "Farm",
"contents": [
(
"@type": "Property",
"name": "id",
"schema": "string"
),
(
"@type": "Relationship",
"@id": "dtmi:com:adt:dtsample:farm:relhaswindmill;1",
"name": "relhaswindmills",
"displayName": "Farm has windmills ",
"target": "dtmi:com:adt:dtsample:floor;1"
)
]
)

(
"@id": "dtmi:com:adt:dtsample:sensor;1",
"@type": "Interface",
"@context": "dtmi:dtdl:context;2",
"displayName": "Sensor",
"contents": [
("@type": "Telemetry", "name": "Temperature", "schema": "double"),
(
"@type": "Property",
"name": "humidity",
"schema": "double"
)
]
)

2.4.2   AMAZON AWS

AWS provides the base cloud hosting solution where different sized server instances (EC2) can be rented to build the digital twin architecture. This IAAS (infrastructure as a service) solution is the basis of cloud hosting. These instances can be bare metal or virtualized instance which can be of different performance and capacity sizes. In addition to hosting on server instances in a particular SLA, popular solutions for data intake and processing of device telemetry for digital twins are Kinesis Data Streams, Kinesis Data Firehose and IoT Core. This ingested data needs to be kept in different storage forms such as relational and Object based, time series and warehouses. DynamoDB and S3 storage are indigenous to AWS. RDS/Aurora, Timestream, Redshift can also be used for data storage. Service hosting environment in AWS like FAAS (function as a service) can be used for data integration and transformation.

Ultimately, in a manner analogous to Azure, a Twin graph, which is a digital thread, can be created in AWS. This digital thread represents a collection of interlinked descriptions of entities—objects, events or concepts. The digital thread puts data in context via linking and semantic metadata.

Consumption and visualization of digital Twin can be done in AWS through a microservice and API Gateway. Traditional monolithic applications are now architected as functional Microservices and hosted as graph. Connectivity and performance SLAs by multiple clients are managed through a service Mesh [200]. A key aspect of digital Twin is security and monitoring. A Hybrid Infrastructure of Edge and Cloud as covered in Section 2.2 and 2.4 can be created in Elastic Kubernetes Service, Elastic Compute Cloud, RedHat OpenShift on AWS. AWS provides a tool to architect the solution well [195].

2.5   BIG DATA

Big data is an important feature of digital twin technology. Chen et al. [45] mention that in contrast to traditional datasets, big data needs more real time analysis. They point out that the IoT involves sensors placed on many different systems which collect and transmit data which needs to be processed in the cloud. At present, big data ranges from several terabytes to several petabytes. Big data has three aspects: volume, velocity and variety. Volume occurs due to the generation and collection of masses of data through sensors or data acquisition systems. Velocity is the rate at which data is received, which depends on the sampling rate. For example, early gas turbine diagnostics involved measurements at a few points during a flight but current systems measure an enormous amount of data in real-time [75]. This large amount of data adds value to the diagnostics process only if it is analyzed properly. For example, raw acceleration data is big data, but when frequencies are extracted from these measurements through FFT, the data becomes easy to analyze. These feature extraction processes are necessary for the successful use of the digital twin. Finally, there is the concept of variety in big data. Data from sensors are typically time series of measurements such as temperature, pressure, strain, deflection, acceleration, frequencies, etc. [53]. However, increasingly, audio and image files are also being routinely uploaded as data. For example, Sarkar et al. [169] used image frames from a video on crack growth for structural health monitoring.

2.5.1 ANALYTICS WITH BIG DATA

With the progress in digital twin development and implementation, a plethora of devices, typically from the manufacturing industry are connected to the IoT. As a result, an enormous amount of data is being generated, sometimes in the order of zetabytes (ZB). Here 1 ZB = 1 billion TB. Big data is typically defined as a large amount of structures, semi-structures and unstructured data emanating from data sources which would need too much expense to store and analyze to extract value. Thus, big data is too massive to be collected, stored or processed by regular data management tools. Big data is often produced in manufacturing during the product life cycle, in activities such as design, manufacturing, maintenance, repair and overhaul (MRO), etc. The strategic significance of big data lies in extracting value through specialized processing. Data visualization can extract features from big data which are more amenable for the use of estimation and machine learning analysis in digital twin design.

Qi and Tao [199] point out the increasing fusion of big data and digital twin technology, especially in the context of smart manufacturing. In the context of MRO, big data obtained from sensors is correlated with historical data such as maintenance data or energy consumption records. Following big data analysis using information processing tools, the product digital twin predicts the health state of the product, the remaining life of the product and the probability of failure.

Hadoop is the computational tool often used in the context of big data. Apache hadoop is an open source framework which can be used to store large data sets in an efficient manner. Furthermore, Hadoop allows processing of large data sets to extract features which may be useful for the digital twin. Instead of using one computer to process the big data set, hadoop uses a cluster of computers in parallel to analyze big data sets. The hadoop ecosystem includes several tools and applications which help to collect, store, process, analyze and manage big data. Some typical applications which are popular are Spark, Presto, Hive, HBase and Zeppelin. Hadoop can be run on AWS platform [194].

2.6 GOOGLE TENSORFLOW

While much of machine learning can be expressed using mathematics, computer implementation needs a high level computer language such as Matlab or Python. Tensorflow can also be considered as a high level language which allows a useful approach for machine learning. Machine learning can be performed with relative ease using high level functions defined Google Tensorflow, an end-to-end open source platform.

TensorFlow provides a comprehensive, flexible ecosystem of tools, libraries and community resources data needs to be preprocessed in order to be useful in training a model. This preprocessing, or "feature engineering", ensures that steps applied during offline training of a model are identical to the steps applied when the model is used to serve predictions. The library tf.Transform in TensorFlow provides consistency of the feature engineering steps during training and serving. It allows users to define pre-processing pipelines and run these using large scale data processing frameworks, while also exporting the pipeline in a way that can be run as part of a TensorFlow graph. Users define a pipeline by composing modular Python functions, which tf.Transform then executes with Apache Beam. The TensorFlow graph exported by tf.Transform enables the preprocessing steps to be replicated when the trained model is used to make predictions, such as when serving the model with TensorFlow Serving (Figure 2.7). Input data: Figure 2.8 provides a description of the raw materials (green) and the settings of the windmill (blue) for illustration.

Example transform of Figure 2.8 using TensorFlow Transform, a library for preprocessing data with (tf.Transform) requires a full-pass, such as:

1. Normalize an input value by mean and standard deviation.
2. Convert strings to integers by generating a vocabulary over all input values.
3. Convert floats to integers by assigning them to buckets based on the observed data distribution.

The output of tf.Transform is exported as a TensorFlow graph (Figures 2.9 and 2.10) to use for training and serving. Code elements for Tensorflow are

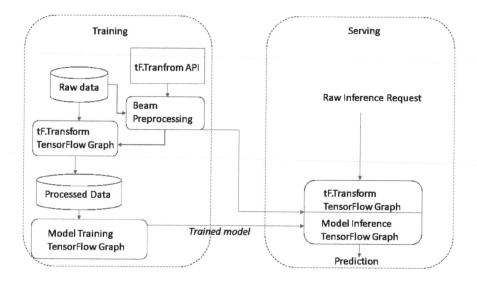

Figure 2.7 Tensorflow training and serving flows.

Blade radius	Tower height	Blade mass	Blade frequency	Blade static stiffness	Blade structural damping	Average wind speed	Standard deviation of wind speed	Rotor material
20	50	50	235	345	12	20	5	Composite
30	100	80	546	237	16	12	3	Metal

Figure 2.8 Description of raw materials and setting of the windmill.

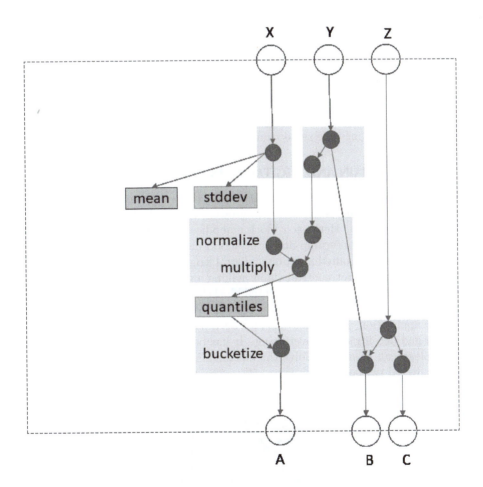

Figure 2.9 Tensorflow graph(a).

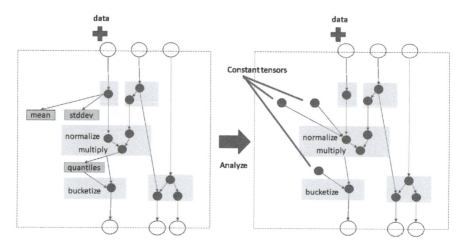

Figure 2.10   Tensorflow graph(b).

shown in Figure 2.11. The same graph is used for both training and serving to prevent skew since the same transformations are applied in both stages.

2.7   BLOCKCHAIN AND DIGITAL TWIN

Huang et al [95] analyzed the digital twin in the context of blockchain technology. A blockchain is defined as a growing list of blocks which are linked using cryptography. Here each block embodies a cryptographic hack of the previous block, a timestamp and transaction data. Blockchain is therefore a decentralized, distributed and public digital ledger. Records in the blockchain cannot be changed retroactively, and making any such change would alter all subsequent blocks. Nakamoto invented blockchain to serve as the public transaction ledger of the cryptocurrency bitcoin. While bitcoin remains its most visible application, blockchain has also been used in banking, privacy preservation and decentralized government. It has become evident that blockchain technology can be used to solve the data management problems which occur in the digital twin of systems.

Figure 2.11   Tensorflow code illustration.

3 Dynamic Systems

In this chapter we discuss dynamics of single and multiple degrees of freedom undamped and viscously damped systems. There are many excellent books [81, 97, 128–130, 138, 144, 153] which discuss these topics very well. Here we give only the details as a necessary background for further developments in the following chapters.

The outline of this chapter is as follows. Dynamic analysis of single-degree-of-freedom undamped systems is discussed in Section 3.1. Single-degree-of-freedom viscously damped systems are considered in Section 3.2. Dynamics of multiple-degree-of-freedom undamped systems are discussed in Section 3.3. The concepts of natural frequencies (eigenvalues) and mode-shapes (eigenvectors) are introduced. Proportionally damped systems are investigated in details in Section 3.4. General non-proportionally damped systems are considered in Section 3.5. Finally, a summary of the topics discussed in this chapter is given in Section 3.6.

3.1 SINGLE-DEGREE-OF-FREEDOM UNDAMPED SYSTEMS

An undamped single-degree-of-freedom (SDOF) system is possibly the simplest system and is of fundamental interest in structural dynamics. A schematic diagram of the system is shown in Figure 3.1.

The equation of motion of the system can be expressed as

$$m\ddot{q}(t) + kq(t) = f(t). \tag{3.1}$$

Here m is the mass, k is the stiffness and $f(t)$ is applied force on the system. The term $\ddot{q}(t)$ denotes the acceleration of the system. The initial conditions associated with the equation of motion can be given by

$$q(0) = q_0 \quad \text{and} \quad \dot{q}(0) = \dot{q}_0. \tag{3.2}$$

It will be shown that the response of the system (3.1) subject to the initial conditions (3.2) can be expressed in terms the natural frequency of the system.

3.1.1 NATURAL FREQUENCY

First consider the harmonic vibration of the system. For this, the forcing function can be expressed as $f(t) = \bar{f}e^{st}$. Assume that the solution $q(t)$ is of the form $q(t) = \bar{q}e^{st}$, $\bar{q}, s \neq 0$. Substituting this in the equation of motion (3.1) as assuming the free vibration (i.e., $\bar{f} = 0$), we obtain

$$\bar{q}\left(ms^2 + k\right) = 0 \quad \text{or} \quad s_{1,2} = \pm i\sqrt{\frac{k}{m}} = \pm i\omega_n. \tag{3.3}$$

DOI: 10.1201/9781003268048-3

35

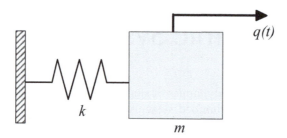

Figure 3.1   A single degree-of-freedom undamped system.

Here i $= \sqrt{-1}$ and w_n, the natural frequency of the system, is given by

$$w_n = \sqrt{\frac{k}{m}}. \tag{3.4}$$

The roots of the characteristic (3.3) are purely imaginary in nature and they appear in complex conjugate pairs. The natural frequency given by Equation 3.4 is the frequency of natural oscillation of the system. The natural time period of the system is given by

$$T_n = \frac{2\pi}{w_n}. \tag{3.5}$$

When the excitation frequency is close to the resonance frequency, we say that the SDOF system is in resonance. Near the resonance frequency, the response of the system can amplify significantly as we will see in the next subsection.

3.1.2   DYNAMIC RESPONSE

We consider the steady-state response of the dynamic system. Taking the Laplace transform of the equation of motion (3.1) and considering the initial conditions in (3.2) one has

$$s^2 m \bar{q}(s) - s m q_0 - m \dot{q}_0 + k \bar{q}(s) = \overline{f}(s) \tag{3.6}$$

$$\text{or} \quad \left(s^2 m + k\right) \bar{q}(s) = \overline{f}(s) + m \dot{q}_0 + s m q_0. \tag{3.7}$$

Here $\bar{q}(s)$ and $\overline{f}(s)$ are Laplace transforms of $q(t)$ and $f(t)$ respectively. The response in the Laplace domain can be obtained as

$$\bar{q}(s) = \frac{\overline{f}(s) + m \dot{q}_0 + s m q_0}{s^2 m + k} = \frac{1}{m} \frac{\overline{f}(s) + m \dot{q}_0 + s m q_0}{s^2 + w_n^2} \tag{3.8}$$

$$= \left\{ \frac{1}{m} \frac{\overline{f}(s)}{s^2 + w_n^2} + \frac{\dot{q}_0}{s^2 + w_n^2} + \frac{s}{s^2 + w_n^2} q_0 \right\}. \tag{3.9}$$

To obtain the vibration response in the time-domain it is required to consider the inverse Laplace transform. Taking the inverse Laplace transform of $\bar{q}(s)$ we have

$$q(t) = \mathcal{L}^{-1}\left[\bar{q}(s)\right] = \mathcal{L}^{-1}\left[\frac{1}{m}\frac{\bar{f}(s)}{s^2 + \omega_n^2}\right] + \mathcal{L}^{-1}\left[\frac{1}{s^2 + \omega_n^2}\right]\dot{q}_0 + \mathcal{L}^{-1}\left[\frac{s}{s^2 + \omega_n^2}\right]q_0.$$
(3.10)

The inverse Laplace transform of the second and third parts can be obtained from the table of Laplace transforms (see for example [107]) as

$$\mathcal{L}^{-1}\left[\frac{1}{s^2 + \omega_n^2}\right] = \frac{\sin\left(\omega_n t\right)}{\omega_n}$$
(3.11)

$$\text{and}\quad \mathcal{L}^{-1}\left[\frac{s}{s^2 + \omega_n^2}\right] = \cos\left(\omega_n t\right).$$
(3.12)

The inverse Laplace transform of the first part can be obtained using the "convolution theorem" [107], which says that the inverse Laplace transform of product of two functions can be expressed as

$$\mathcal{L}^{-1}\left[\bar{f}(s)\bar{g}(s)\right] = \int_0^t f(\tau)g(t - \tau)d\tau.$$
(3.13)

Considering $\bar{g}(s) = \frac{1}{m}\frac{1}{s^2 + \omega_n^2}$, the inverse Laplace transform of the first part can be obtained as

$$\mathcal{L}^{-1}\left[\bar{f}(s)\frac{1}{s^2 + \omega_n^2}\right] = \int_0^t \frac{1}{m\omega_n}f(\tau)\sin\left(\omega_n(t - \tau)\right)d\tau.$$
(3.14)

Combining Equations 3.14, 3.11 and 3.12, from Equation 3.10 we have

$$q(t) = \int_0^t \frac{1}{m\omega_n}f(\tau)\sin\left(\omega_n(t - \tau)\right)d\tau + \frac{1}{\omega_n}\sin(\omega_n t)\dot{q}_0 + \cos(\omega_n t)q_0.$$
(3.15)

Collecting the terms associated with $\sin\left(\omega_n t\right)$ and $\cos\left(\omega_n t\right)$ this expression can be simplified as

$$q(t) = \frac{1}{m\omega_n}\int_0^t f(\tau)\sin\left(\omega_n(t - \tau)\right)d\tau + B\cos\left(\omega_n t - \vartheta\right)$$
(3.16)

where the amplitude and the phase angle can be obtained from

$$B = \sqrt{q_0^2 + \left(\frac{\dot{q}_0}{\omega_n}\right)^2}$$
(3.17)

$$\text{and}\quad \tan\vartheta = \frac{\dot{q}_0}{\omega_n q_0}.$$
(3.18)

The second part of Equation 3.16, i.e., the term $B\cos\left(\omega_n t - \vartheta\right)$ only depends on the initial conditions and independent of the applied forcing. The first part on the other hand only depends on the applied forcing. The complete response of the SDOF system is given by Equation 3.15.

3.1.2.1 Impulse Response Function

In the special case when all the initial conditions are zero and the forcing function is a Dirac delta function, that is $f(\tau) = \delta(\tau)$, the response is known as the impulse response function. This can be obtained from Equation 3.15 by substituting $q_0 = 0, \dot{q}_0 = 0$ and $f(\tau) = \delta(\tau)$ as

$$h(t) = \int_0^t \frac{1}{m\omega_n} \delta(\tau) \sin\left(\omega_n(t - \tau)\right) d\tau = \frac{1}{m\omega_n} \sin \omega_n t. \quad (3.19)$$

This is a fundamental property of the system. If this function is known (e.g., from experiments), then the response to any forcing can be obtained from the convolution integral expressions in (3.15) as

$$q(t) = \int_0^t f(\tau) h(t - \tau) d\tau. \quad (3.20)$$

Next we consider viscously damped single-degree-of-freedom systems.

3.2 SINGLE-DEGREE-OF-FREEDOM VISCOUSLY DAMPED SYSTEMS

A single-degree-of-freedom (SDOF) system with viscous damping is shown in Figure 3.2. The equation of motion of the system can be expressed as

$$m\ddot{q}(t) + c\dot{q}(t) + kq(t) = f(t). \quad (3.21)$$

Here m is the mass of the system, c is the damping, k is the spring stiffness and $f(t)$ is the applied forcing. As before, the symbol ($\bullet$) represents derivative with respect to time, so that $\dot{q}(t)$ denotes the velocity and $\ddot{q}(t)$ denotes the acceleration of the system. The initial conditions associated with the equation of motion can be given by (3.2). It will be shown that the response of the system (3.21) subject to the initial conditions (3.2) can be expressed in terms the undamped and damped natural frequencies of the system.

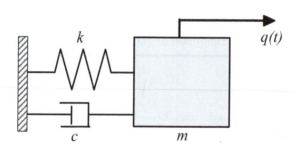

Figure 3.2 A single degree-of-freedom system with viscous damping.

3.2.1 NATURAL FREQUENCY

Like the undamped system, first we consider harmonic vibration of the system. For this, both the forcing function and the solution can be expressed as $f(t) = \bar{f}e^{st}$ and $q(t) = \bar{q}e^{st}$, $\bar{q}, s \neq 0$. Substituting these in the equation of motion (3.21) and assuming the free vibration (i.e., $\bar{f} = 0$), we obtain the characteristic equation

$$\bar{q}\left(ms^2 + +sc + k\right) = 0 \quad \text{or} \quad \frac{\bar{q}}{m}\left(s^2 + +s\frac{c}{m} + \frac{k}{m}\right) = 0. \tag{3.22}$$

We introduce the viscous damping factor as

$$\zeta = \frac{c}{2\sqrt{k\,m}}. \tag{3.23}$$

Using the definition of the natural frequency in Equation 3.4 it can be shown that

$$\frac{c}{m} = 2\zeta\omega_n \tag{3.24}$$

Using this, from Equation 3.22 we have the characteristic equation

$$s^2 + +2\zeta\omega_n\,s + \omega_n^2 = 0 \quad \text{or} \quad s_{1,2} = -\zeta\omega_n + \pm i\omega_n\sqrt{1 - \zeta^2}. \tag{3.25}$$

The nature of the dynamic response of the system depends on the characteristics of the two roots in (3.25). It turns out that the value of ζ controls the characteristics of both the roots. It is useful to define the critical damping factor

$$c_{cr} = 2\sqrt{km} \tag{3.26}$$

so that the the damping factor in (3.23) becomes the ratio

$$\zeta = \frac{c}{c_{cr}}. \tag{3.27}$$

Three possible cases arise based on the value of ζ:

1. Overdamped motion, $\zeta > 1$: In this case both roots in (3.25) are real and negative as

$$s_1 = -\omega_n\left(\zeta + \sqrt{\zeta^2 - 1}\right) \quad \text{and} \quad s_2 = -\omega_n\left(\zeta - \sqrt{\zeta^2 - 1}\right). \tag{3.28}$$

   The motion of the system is therefore nonoscillatory and exponentially decaying in nature.
2. Critically damped motion, $\zeta = 1$: In this case both roots in (3.25) are real and negative and identical in nature. That is, the system have repeated roots as

$$s_1 = -\omega_n \quad \text{and} \quad s_2 = -\omega_n. \tag{3.29}$$

   The motion of the system can be expressed as $q(t) = (a_1 + a_2t)e^{-\omega_n t}$ for some real constants a_1 and a_2. This motion is also nonoscillatory and exponentially decaying in nature. The motion of a critically damped system can be viewed as the borderline between nonoscillatory and oscillatory motion.

3. Underdamped motion, $\zeta < 1$: In this case the roots of the characteristic Equation 3.25 appear in complex conjugate pairs as

$$s_1 = -\zeta\omega_n + i\omega_d \quad \text{and} \quad s_2 = -\zeta\omega_n - i\omega_d. \tag{3.30}$$

Here the damped natural frequency of the system, is given by

$$\omega_d = \omega_n\sqrt{1-\zeta^2} = \sqrt{\frac{k}{m}}\sqrt{1-\zeta^2}. \tag{3.31}$$

Consequently, the damped time period of the system can be defined as

$$T_d = \frac{2\pi}{\omega_d}. \tag{3.32}$$

In view of the roots in (3.30), the motion of the system can be expressed as $q(t) = Be^{-\zeta\omega_n t}\sin(\omega_d t\vartheta)$. The motion of the system is therefore oscillatory and exponentially decaying in nature.

3.2.2 DYNAMIC RESPONSE

We consider steady-sate response of the dynamic system. Taking the Laplace transform of the equation of motion (3.21) and considering the initial conditions in (3.2) one has

$$s^2 m\bar{q}(s) - smq_0 - m\dot{q}_0 + sc\bar{q}(s) - cq_0 + k\bar{q}(s) = \bar{f}(s) \tag{3.33}$$

$$\text{or} \quad \left(s^2 m + sc + k\right)\bar{q}(s) = \bar{f}(s) + m\dot{q}_0 + smq_0 + cq_0. \tag{3.34}$$

The response in the Laplace domain can be obtained as

$$\bar{q}(s) = \frac{\bar{f}(s) + m\dot{q}_0 + smq_0 + cq_0}{s^2 m + sc + k} = \frac{1}{m}\frac{\bar{f}(s) + m\dot{q}_0 + smq_0 + cq_0}{s^2 + 2s\zeta\omega_n + \omega_n^2} \tag{3.35}$$

$$= \left\{\frac{1}{m}\frac{\bar{f}(s)}{s^2 + 2s\zeta\omega_n + \omega_n^2} + \frac{\dot{q}_0 + 2\zeta\omega_n q_0}{s^2 + 2s\zeta\omega_n + \omega_n^2} + \frac{s}{s^2 + 2s\zeta\omega_n + \omega_n^2}q_0\right\}. \tag{3.36}$$

Using the definition of the damped natural frequency from Equation 3.31 we can reorganize the denominator as

$$s^2 + 2s\zeta\omega_n + \omega_n^2 = (s + \zeta\omega_n)^2 - (\zeta\omega_n)^2 + \omega_n^2 = (s + \zeta\omega_n)^2 + \omega_d^2. \tag{3.37}$$

To obtain the vibration response in the time-domain, it is required to consider the inverse Laplace transform. Taking the inverse Laplace transform of

$\bar{q}(s)$ in (3.36) we have

$$q(t) = \mathcal{L}^{-1}\left[\bar{q}(s)\right] = \mathcal{L}^{-1}\left[\frac{1}{m}\frac{\bar{f}(s)}{(s+\zeta\omega_n)^2 + \omega_d^2}\right] +$$

$$\mathcal{L}^{-1}\left[\frac{1}{(s+\zeta\omega_n)^2 + \omega_d^2}\right](\dot{q}_0 + 2\zeta\omega_n q_0)$$

$$+ \mathcal{L}^{-1}\left[\frac{s}{(s+\zeta\omega_n)^2 + \omega_d^2}\right]q_0. \quad (3.38)$$

The inverse Laplace transform of the second and third parts can be obtained from the table of Laplace transforms (see for example [107]) as

$$\mathcal{L}^{-1}\left[\frac{1}{(s+\alpha)^2 + \beta^2}\right] = \frac{e^{-\alpha t}\sin(\beta t)}{\beta} \quad (3.39)$$

$$\text{and} \quad \mathcal{L}^{-1}\left[\frac{s}{(s+\alpha)^2 + \beta^2}\right] = e^{-\alpha t}\cos(\beta t) - \frac{\alpha e^{-\alpha t}\sin(\beta t)}{\beta}. \quad (3.40)$$

Here α and β stand for $\zeta\omega_n$ and ω_d, respectively.

The inverse Laplace transform of the first part can be obtained using the convolution theorem (3.13). Considering $\bar{g}(s) = \frac{1}{m}\frac{1}{(s+\zeta\omega_n)^2 + \omega_d^2}$, the inverse Laplace transform of the first part can be obtained as

$$\mathcal{L}^{-1}\left[\bar{f}(s)\frac{1}{(s+\zeta\omega_n)^2 + \omega_d^2}\right] = \int_0^t \frac{1}{m\omega_d}f(\tau)e^{-\zeta\omega_n(t-\tau)}\sin(\omega_d(t-\tau))\,d\tau. \quad (3.41)$$

Combining (3.39), (3.40) and (3.41), from Equation 3.38 we have

$$q(t) = \int_0^t \frac{1}{m\omega_d}f(\tau)e^{-\zeta\omega_n(t-\tau)}\sin(\omega_d(t-\tau))\,d\tau +$$

$$\frac{e^{-\zeta\omega_n t}}{\omega_d}\sin(\omega_d t)(\dot{q}_0 + 2\zeta\omega_n q_0) +$$

$$\left\{e^{-\zeta\omega_n t}\cos(\omega_d t) - \frac{\zeta\omega_n e^{-\zeta\omega_n t}\sin(\omega_d t)}{\omega_d}\right\} \quad (3.42)$$

$$q_0 = \int_0^t \frac{1}{m\omega_d}f(\tau)e^{-\zeta\omega_n(t-\tau)}\sin(\omega_d(t-\tau))\,d\tau +$$

$$\sin(\omega_n t)\left(\frac{\dot{q}_0 + \zeta\omega_n q_0}{\omega_d}\right) +$$

$$\cos(\omega_n t)q_0.$$

Collecting the terms associated with $\sin(\omega_d t)$ and $\cos(\omega_d t)$ this expression can be simplified as

$$q(t) = \int_0^t \frac{1}{m\omega_d}f(\tau)e^{-\zeta\omega_n(t-\tau)}\sin(\omega_d(t-\tau))\,d\tau + B\cos(\omega_d t - \vartheta) \quad (3.43)$$

where the amplitude and the phase angle can be obtained from

$$B = \sqrt{q_0^2 + \left(\frac{\dot{q}_0 + \zeta \omega_n q_0}{\omega_d} \right)^2} \qquad (3.44)$$

$$\text{and} \quad \tan \vartheta = \frac{\dot{q}_0 + \zeta \omega_n q_0}{\omega_d q_0}. \qquad (3.45)$$

The second part of Equation 3.43, i.e., the term $B \cos (\omega_n t - \vartheta)$ only depends on the initial conditions and independent of the applied forcing. The first part on the other hand only depends on the applied forcing. The complete response of a viscously damped SDOF system is given by Equation 3.42. In the special case when the system is undamped, by substituting $\zeta = 0$ one can obtain the results derived in the previous section.

3.2.2.1 Impulse Response and Frequency Response Function

When all the initial conditions are zero and the forcing function is a Dirac delta function, that is $f(\tau) = \delta(\tau)$, the response is known as the impulse response function. This can be obtained from Equation 3.42 by substituting $q_0 = 0, \dot{q}_0 = 0$ and $f(\tau) = \delta(\tau)$ as

$$h(t) = \int_0^t \frac{1}{m\omega_d} \delta(\tau) e^{-\zeta \omega_n (t-\tau)} \sin (\omega_d(t - \tau)) \, d\tau = \frac{1}{m\omega_d} e^{-\zeta \omega_n t} \sin \omega_n t.$$
$$(3.46)$$

This is a fundamental property of the damped system and depends only on the natural frequency and the damping factor. If this function is known (e.g., from experiments), then the response to any forcing can be obtained from the convolution integral expressions in (3.42) as $q(t) = \int_0^t f(\tau)h(t - \tau)d\tau$.

The frequency response function can be obtained directly from Equation 3.34 by substituting $q_0 = 0, \dot{q}_0 = 0, \overline{f}(s) = 1$ and $s = i\omega$ as

$$\bar{h}(i\omega) = \frac{1}{-\omega^2 m + i\omega c + k} = \frac{1}{m} \frac{1}{(-\omega^2 + 2i\zeta \omega \omega_n + \omega_n^2)}. \qquad (3.47)$$

Here ω is the driving frequency.

For convenience we introduce a normalized time $t' = t/T_n = \omega_n t/2\pi$. Using this, the normalized impulse response function is shown in Figure 3.3(a) for four values of the damping factor, namely $\zeta = 0.01, 0.1, 0.25$ and 0.5. It can be seen that with increasing damping (i) the oscillations diminish to almost zero very quickly, and (ii) the amplitude of successive peaks also reduces significantly. For example when $\zeta = 0.5$, the vibration ceases after about three period of oscillations.

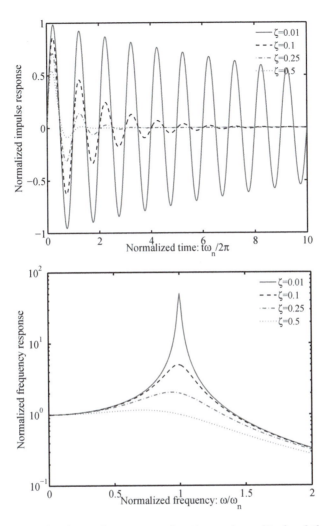

Figure 3.3 Normalized impulse response function and amplitude of the frequency response function of a viscously damped SDOF system for various values of the damping factor ζ.

Dividing the expression for the frequency response function in (3.47) by ω_n and introducing the nondimensional frequency parameter $\tilde{\omega} = \omega/\omega_n$ we have

$$\bar{h}(i\omega) = \frac{1}{m} \frac{1}{(-\omega^2 + 2i\zeta\omega\omega_n + \omega_n^2)}$$

$$= \frac{1}{m\omega_n^2} \frac{1}{(-\tilde{\omega}^2 + 2i\zeta\tilde{\omega} + 1)} = \frac{1}{k} \frac{1}{(-\tilde{\omega}^2 + 2i\zeta\tilde{\omega} + 1)}. \qquad (3.48)$$

The amplitude of the normalized frequency response function $(\bar{h}(i\tilde{\omega})/(1/k))$ can be obtained as

$$\left|\frac{\bar{h}(i\tilde{\omega})}{(1/k)}\right| = \frac{1}{\sqrt{(1-\tilde{\omega}^2)^2 + 4\zeta^2\tilde{\omega}^2}}. \tag{3.49}$$

This function is shown in Figure 3.3(b) for the same four values of the damping factor considered to plot the impulse response function. The following observations can be made:

For lower values of the damping factor, the response amplitude has a peak near $\omega = \omega_n$

The higher the damping, the lower the value of the peak response

The frequency of the peak response shifts to the left for increasing damping values

These observations can be explained by analytically obtaining the 'peak frequency'. Differentiating the amplitude of the normalized frequency response function with respect to $\tilde{\omega}$ and setting it to zero we have

$$\frac{d}{d\tilde{\omega}}\left|\frac{\bar{h}(i\tilde{\omega})}{(1/k)}\right| = 0 \quad \text{or} \quad \frac{d}{d\tilde{\omega}}\left[(1-\tilde{\omega}^2)^2 + 4\zeta^2\tilde{\omega}^2\right]^{-1/2}. \tag{3.50}$$

Solving this equation we have

$$\tilde{\omega}_{\max} = \frac{\omega_{\max}}{\omega_n} = \sqrt{1 - 2\zeta^2}. \tag{3.51}$$

For real values of $\tilde{\omega}_{\max}$, one must have $1 - 2\zeta^2 \geq 0$, or $\zeta < 1/\sqrt{2}$. Provided the value of the damping factor is lower than $1/\sqrt{2} = 0.7071$, there will be a peak in the frequency response function. The maximum amplitude can be obtained by substituting the $\tilde{\omega}_{\max}$ in the expression of the normalized frequency response as (3.49) as

$$\left|\frac{\bar{h}(i\tilde{\omega})}{(1/k)}\right|_{\max} = \frac{1}{2\zeta\sqrt{1-\zeta^2}}. \tag{3.52}$$

This shows that the maximum amplitude will decrease with increasing value of the damping factor. The influence of the damping factor on the dynamics of a viscously damped single-degree-of-freedom system is summarized in 3.1.

3.3   MULTIPLE-DEGREE-OF-FREEDOM UNDAMPED SYSTEMS

The equation of motion of an undamped non-gyroscopic system with N degrees of freedom can be given by

$$\mathbf{M\ddot{q}}(t) + \mathbf{Kq}(t) = \mathbf{f}(t) \tag{3.53}$$

Table 3.1

The Influence of the Damping Factor on the Dynamics of a Viscously Damped Single-Degree-of-Freedom System

Parameter values	Dynamic characteristics
$\zeta > 1$	Overdamped motion—the motion is nonoscillatory and exponentially decaying
$\zeta = 1$	Critically damped motion—the motion is nonoscillatory and exponentially decaying
$\zeta < 1$	Underdamped motion—the motion is oscillatory and exponentially decaying
$1 < \zeta \leq 1/\sqrt{2}$	Oscillatory motion, but no peak in the frequency response function
$1/\sqrt{2} < \zeta$	Oscillatory motion, the frequency response function has peak at a normalized frequency $\sqrt{1 - 2\zeta^2}$ with a normalized peak value of $\dfrac{1}{2\zeta\sqrt{1-\zeta^2}}$

where $\mathbf{M} \in \mathbb{R}^{N \times N}$ is the mass matrix, $\mathbf{K} \in \mathbb{R}^{N \times N}$ is the stiffness matrix, $\mathbf{q}(t) \in \mathbb{R}^{N}$ is the vector of generalized coordinates and $\mathbf{f}(t) \in \mathbb{R}^{N}$ is the forcing vector. Equation 3.53 represents a set of coupled second-order ordinary-differential equations. The solution of this equation also requires the knowledge of the initial conditions in terms of displacements and velocities of all the coordinates. The initial conditions can be specified as

$$\mathbf{q}(0) = \mathbf{q}_0 \in \mathbb{R}^{N} \quad \text{and} \quad \dot{\mathbf{q}}(0) = \dot{\mathbf{q}}_0 \in \mathbb{R}^{N}. \tag{3.54}$$

We aim to solve Equation 3.53 together with the initial conditions in 3.54 using modal analysis.

3.3.1 MODAL ANALYSIS

Lord Rayleigh [157] showed that undamped linear systems, equations of motion of which are given by (3.53), are capable of so-called *natural motions*. This essentially implies that all the system coordinates execute harmonic oscillation at a given frequency and form a certain displacement pattern. The oscillation frequency and displacement pattern are called *natural frequencies* and *normal modes*, respectively. The natural frequencies (ω_j) and the mode shapes ($\mathbf{x}_j$) are intrinsic characteristic of a system and can be obtained by solving the associated matrix eigenvalue problem

$$\mathbf{K}\mathbf{x}_j = \omega_j^2 \mathbf{M}\mathbf{x}_j, \quad \forall j = 1, \cdots, N. \tag{3.55}$$

Since the above eigenvalue problem is in terms of real symmetric nonnegative definite matrices $\mathbf{M}$ and $\mathbf{K}$, the eigenvalues and consequently the eigenvectors are real, that is $\omega_j \in \mathbb{R}$ and $\mathbf{x}_j \in \mathbb{R}^N$. Premultiplying Equation 3.55 by $\mathbf{x}_k^T$ we have

$$\mathbf{x}_k^T \mathbf{K} \mathbf{x}_j = \omega_j^2 \mathbf{x}_k^T \mathbf{M} \mathbf{x}_j. \tag{3.56}$$

Taking transpose of the above equation and noting that $\mathbf{M}$ and $\mathbf{K}$ are symmetric matrices one has

$$\mathbf{x}_j^T \mathbf{K} \mathbf{x}_k = \omega_j^2 \mathbf{x}_j^T \mathbf{M} \mathbf{x}_k. \tag{3.57}$$

Now consider the eigenvalue equation for the kth mode:

$$\mathbf{K} \mathbf{x}_k = \omega_k^2 \mathbf{M} \mathbf{x}_k. \tag{3.58}$$

Premultiplying Equation 3.58 by $\mathbf{x}_j^T$ we have

$$\mathbf{x}_j^T \mathbf{K} \mathbf{x}_k = \omega_k^2 \mathbf{x}_j^T \mathbf{M} \mathbf{x}_k. \tag{3.59}$$

Subtracting Equation 3.57 from (3.59) we have

$$\left(\omega_k^2 - \omega_j^2\right) \mathbf{x}_j^T \mathbf{M} \mathbf{x}_k = 0. \tag{3.60}$$

Since we assumed that the natural frequencies are not repeated, when $j \neq k$, $\omega_j \neq \omega_k$. Therefore, from Equation 3.60 it follows that

$$\mathbf{x}_k^T \mathbf{M} \mathbf{x}_j = 0. \tag{3.61}$$

Using this in Equation 3.57 we can also obtain

$$\mathbf{x}_k^T \mathbf{K} \mathbf{x}_j = 0. \tag{3.62}$$

These two relationships often known as the mode orthogonality relationships. If we normalize $\mathbf{x}_j$ such that $\mathbf{x}_j \mathbf{M} \mathbf{x}_j = 1$, then from Equation 3.57 it follows that $\mathbf{x}_j \mathbf{K} \mathbf{x}_j = \omega_j^2$. This normalization is known as the unity mass normalization, a convention often used in practice. Equations (3.61) and (3.62) are known as the orthogonality relationships. These equations combined with the normalization relationships can be concisely written in terms of Kroneker delta function δ_{lj} as

$$\mathbf{x}_l^T \mathbf{M} \mathbf{x}_j = \delta_{lj} \tag{3.63}$$

$$\text{and} \quad \mathbf{x}_l^T \mathbf{K} \mathbf{x}_j = \omega_j^2 \delta_{lj}, \quad \forall\, l, j = 1, \cdots, N. \tag{3.64}$$

Note that $\delta_{lj} = 1$ for $l = j$ and $\delta_{lj} = 0$ otherwise. The property of the eigenvectors in (3.63) is also known as the mass orthonormality relationship. The solution of undamped eigenvalue problem is now standard in many finite element packages. There are various efficient algorithms available for this purpose, see for example [14, 207]. We refer to these publications [11, 12, 49] for Krylov subspace based techniques for further numerical efficiency.

This orthogonality property of the undamped modes is very powerful as it allows to transform a set of *coupled* differential equations to a set of *independent* equations. For convenience, we construct the matrices

$$\mathbf{\Omega} = \text{diag}\,[\omega_1, \omega_2, \cdots, \omega_N] \in \mathbb{R}^{N \times N} \tag{3.65}$$

$$\text{and}\quad \mathbf{X} = [\mathbf{x}_1, \mathbf{x}_2, \cdots, \mathbf{x}_N] \in \mathbb{R}^{N \times N} \tag{3.66}$$

where the eigenvalues are arranged such that $\omega_1 < \omega_2$, $\omega_2 < \omega_3, \cdots, \omega_k < \omega_{k+1}$. The matrix $\mathbf{X}$ is knows as the undamped modal matrix. Using these matrix notations, the orthogonality relationships (3.63) and (3.64) can be rewritten as

$$\mathbf{X}^T \mathbf{M} \mathbf{X} = \mathbf{I} \tag{3.67}$$

$$\text{and}\quad \mathbf{X}^T \mathbf{K} \mathbf{X} = \mathbf{\Omega}^2 \tag{3.68}$$

where $\mathbf{I}$ is a $N \times N$ identity matrix. We use the following coordinate transformation (known as the modal transformation)

$$\mathbf{q}(t) = \mathbf{X}\mathbf{y}(t). \tag{3.69}$$

Substituting $\mathbf{q}(t)$ in Equation 3.53, premultiplying by $\mathbf{X}^T$ and using the orthogonality relationships in (3.67) and (3.68), the equation of motion in the modal coordinates may be obtained as

$$\begin{aligned} \ddot{\mathbf{y}}(t) + \mathbf{\Omega}^2 \mathbf{y}(t) &= \mathbf{f}'(t) \\ \text{or}\quad \ddot{y}_j(t) + \omega_j^2 y_j(t) &= f'_j(t) \quad \forall j = 1, \cdots, N \end{aligned} \tag{3.70}$$

where $\bar{\mathbf{f}}'(t) = \mathbf{X}^T \mathbf{f}(t)$ is the forcing function in the modal coordinates. This method significantly simplifies the dynamic analysis because complex multiple degrees of freedom systems can be treated as collections of single-degree-of-freedom oscillators. This approach of analyzing linear undamped systems is known as the *modal analysis*, possibly the most efficient tool for vibration analysis of complex engineering structures. Beside the numerical efficiency, the natural frequencies and the mode shapes provide valuable physical insights into the nature of vibration of the system.

3.3.2 DYNAMIC RESPONSE

In this subsection we outline how modal analysis can be used to obtain expressions for dynamic response of a system. Analysis in the frequency domain is described first, followed by the time-domain analysis.

3.3.2.1 Frequency-Domain Analysis

We consider steady-sate response of the dynamic system. Taking the Laplace transform of (3.53) and considering the initial conditions in (3.54) one has

$$s^2 \mathbf{M}\bar{\mathbf{q}}(s) - s\mathbf{M}\mathbf{q}_0 - \mathbf{M}\dot{\mathbf{q}}_0 + \mathbf{K}\bar{\mathbf{q}}(s) = \bar{\mathbf{f}}(s) \tag{3.71}$$

$$\text{or}\quad \left[s^2 \mathbf{M} + \mathbf{K}\right]\bar{\mathbf{q}}(s) = \bar{\mathbf{f}}(s) + \mathbf{M}\dot{\mathbf{q}}_0 + s\mathbf{M}\mathbf{q}_0 = \bar{\mathbf{p}}(s) \;\; \text{(say)}. \tag{3.72}$$

Using the modal transformation

$$\bar{\mathbf{q}}(s) = \mathbf{X}\bar{\mathbf{y}}(s) \tag{3.73}$$

and premultiplying (3.72) by $\mathbf{X}^T$, we have

$$\left[s^2\mathbf{M} + \mathbf{K}\right]\mathbf{X}\bar{\mathbf{y}}(s) = \bar{\mathbf{p}}(s) \quad \text{or} \quad \left\{\mathbf{X}^T\left[s^2\mathbf{M} + \mathbf{K}\right]\mathbf{X}\right\}\bar{\mathbf{y}}(s) = \mathbf{X}^T\bar{\mathbf{p}}(s). \tag{3.74}$$

Using the orthogonality relationships in (3.67) and (3.68), this equation reduces to

$$\left[s^2\mathbf{I} + \boldsymbol{\Omega}^2\right]\bar{\mathbf{y}}(s) = \mathbf{X}^T\bar{\mathbf{p}}(s) \tag{3.75}$$

$$\text{or} \quad \bar{\mathbf{y}}(s) = \left[s^2\mathbf{I} + \boldsymbol{\Omega}^2\right]^{-1}\mathbf{X}^T\bar{\mathbf{p}}(s) \tag{3.76}$$

$$\text{or} \quad \mathbf{X}\bar{\mathbf{y}}(s) = \mathbf{X}\left[s^2\mathbf{I} + \boldsymbol{\Omega}^2\right]^{-1}\mathbf{X}^T\bar{\mathbf{p}}(s) \quad \text{(premultiplying by } \mathbf{X}\text{)} \tag{3.77}$$

$$\text{or} \quad \bar{\mathbf{q}}(s) = \mathbf{X}\left[s^2\mathbf{I} + \boldsymbol{\Omega}^2\right]^{-1}\mathbf{X}^T\bar{\mathbf{p}}(s) \quad \text{(using 3.73)} \tag{3.78}$$

$$\text{or} \quad \bar{\mathbf{q}}(s) = \mathbf{X}\left[s^2\mathbf{I} + \boldsymbol{\Omega}^2\right]^{-1}\mathbf{X}^T\left\{\bar{\mathbf{f}}(s) + \mathbf{M}\dot{\mathbf{q}}_0 + s\mathbf{M}\mathbf{q}_0\right\} \quad \text{(using 3.72)}. \tag{3.79}$$

Equation 3.79 is the complete solution of the undamped dynamic response using modal analysis. In structural dynamics often frequency-domain analysis is used. The dynamic response in the frequency domain can be obtained by substituting $s = i\omega$ as

$$\begin{aligned}\bar{\mathbf{q}}(i\omega) &= \mathbf{X}\left[-\omega^2\mathbf{I} + \boldsymbol{\Omega}^2\right]^{-1}\mathbf{X}^T\left\{\bar{\mathbf{f}}(i\omega) + \mathbf{M}\dot{\mathbf{q}}_0 + i\omega\mathbf{M}\mathbf{q}_0\right\} \\ &= \mathbf{H}(i\omega)\left\{\bar{\mathbf{f}}(i\omega) + \mathbf{M}\dot{\mathbf{q}}_0 + i\omega\mathbf{M}\mathbf{q}_0\right\}.\end{aligned} \tag{3.80}$$

The term appearing in the preceding equation

$$\mathbf{H}(i\omega) = \mathbf{X}\left[-\omega^2\mathbf{I} + \boldsymbol{\Omega}^2\right]^{-1}\mathbf{X}^T \tag{3.81}$$

is often known as the transfer function matrix. Note that $\left[-\omega^2\mathbf{I} + \boldsymbol{\Omega}^2\right]$ is a diagonal matrix and therefore its inverse is easy to obtain since

$$\left[-\omega^2\mathbf{I} + \boldsymbol{\Omega}^2\right]^{-1} = \text{diag}\left[\frac{1}{\omega_1^2 - \omega^2}, \frac{1}{\omega_2^2 - \omega^2}, \cdots, \frac{1}{\omega_N^2 - \omega^2}\right]. \tag{3.82}$$

The product $\mathbf{X}\left[-\omega^2\mathbf{I}+\mathbf{\Omega}^2\right]^{-1}\mathbf{X}^T$ can be expressed as

$$\mathbf{X}\left[-\omega^2\mathbf{I}+\mathbf{\Omega}^2\right]^{-1}\mathbf{X}^T \qquad (3.83)$$

$$= [\mathbf{x}_1, \mathbf{x}_2, \cdots, \mathbf{x}_N]\,\mathrm{diag}\left[\frac{1}{\omega_1^2-\omega^2}, \frac{1}{\omega_2^2-\omega^2}, \cdots, \frac{1}{\omega_N^2-\omega^2}\right]\begin{bmatrix}\mathbf{x}_1^T\\\mathbf{x}_2^T\\\vdots\\\mathbf{x}_N^T\end{bmatrix}$$

$$= [\mathbf{x}_1, \mathbf{x}_2, \cdots, \mathbf{x}_N]\begin{bmatrix}\frac{\mathbf{x}_1^T}{\omega_1^2-\omega^2}\\\frac{\mathbf{x}_2^T}{\omega_2^2-\omega^2}\\\vdots\\\frac{\mathbf{x}_N^T}{\omega_N^2-\omega^2}\end{bmatrix} \qquad (3.84)$$

$$= \left[\frac{\mathbf{x}_1\mathbf{x}_1^T}{\omega_1^2-\omega^2} + \frac{\mathbf{x}_2\mathbf{x}_2^T}{\omega_2^2-\omega^2} + \cdots + \frac{\mathbf{x}_N\mathbf{x}_N^T}{\omega_N^2-\omega^2}\right].$$

From this we obtain the familiar expression of the transfer function matrix as

$$\mathbf{H}(\mathrm{i}\omega) = \sum_{j=1}^{N}\frac{\mathbf{x}_j\mathbf{x}_j^T}{\omega_j^2-\omega^2}. \qquad (3.85)$$

Substituting $\mathbf{H}(\mathrm{i}\omega)$ in the expression of dynamic response (3.80) we have

$$\bar{\mathbf{q}}(\mathrm{i}\omega) = \sum_{j=1}^{N}\frac{\mathbf{x}_j\mathbf{x}_j^T\left\{\bar{\mathbf{f}}(\mathrm{i}\omega)+\mathbf{M}\dot{\mathbf{q}}_0+\mathrm{i}\omega\mathbf{M}\mathbf{q}_0\right\}}{\omega_j^2-\omega^2} = \qquad (3.86)$$

$$\sum_{j=1}^{N}\frac{\mathbf{x}_j^T\bar{\mathbf{f}}(\mathrm{i}\omega)+\mathbf{x}_j^T\mathbf{M}\dot{\mathbf{q}}_0+\mathrm{i}\omega\mathbf{x}_j^T\mathbf{M}\mathbf{q}_0}{\omega_j^2-\omega^2}\mathbf{x}_j.$$

This expression shows that the dynamic response of the system is a linear combination of the mode shapes.

3.3.2.2 Time-Domain Analysis

Rewriting the expression of the frequency-domain response in Equation 3.86 in the Laplace domain we have

$$\bar{\mathbf{q}}(s) = \sum_{j=1}^{N}\left\{\frac{\mathbf{x}_j^T\bar{\mathbf{f}}(s)}{s^2+\omega_j^2} + \frac{\mathbf{x}_j^T\mathbf{M}\dot{\mathbf{q}}_0}{s^2+\omega_j^2} + \frac{s}{s^2+\omega_j^2}\mathbf{x}_j^T\mathbf{M}\mathbf{q}_0\right\}\mathbf{x}_j. \qquad (3.87)$$

To obtain the vibration response in the time domain it is required to consider the inverse Laplace transform. Taking the inverse Laplace transform of $\bar{\mathbf{q}}(s)$

we have

$$\mathbf{q}(t) = \mathcal{L}^{-1}\left[\bar{\mathbf{q}}(s)\right] = \sum_{j=1}^{N} a_j(t)\mathbf{x}_j \tag{3.88}$$

where the time dependent constants are given by

$$a_j(t) = \mathcal{L}^{-1}\left[\frac{\mathbf{x}_j^T\bar{\mathbf{f}}(s)}{s^2 + \omega_j^2}\right] + \mathcal{L}^{-1}\left[\frac{1}{s^2 + \omega_j^2}\right]\mathbf{x}_j^T\mathbf{M}\dot{\mathbf{q}}_0 + \mathcal{L}^{-1}\left[\frac{s}{s^2 + \omega_j^2}\right]\mathbf{x}_j^T\mathbf{M}\mathbf{q}_0. \tag{3.89}$$

Following the inverse Laplace transform approach described in Section 3.1.2, the time dependent coefficients a_j in Equation 3.89 can be expressed as

$$a_j(t) = \int_0^t \frac{1}{\omega_j}\mathbf{x}_j^T\mathbf{f}(\tau)\sin\left(\omega_j(t-\tau)\right)d\tau + \frac{1}{\omega_j}\sin(\omega_j t)\mathbf{x}_j^T\mathbf{M}\dot{\mathbf{q}}_0 + \cos(\omega_j t)\mathbf{x}_j^T\mathbf{M}\mathbf{q}_0. \tag{3.90}$$

Collecting the terms associated with $\sin\left(\omega_{d_j} t\right)$ and $\cos\left(\omega_{d_j} t\right)$ this expression can be simplified as

$$a_j(t) = \int_0^t \frac{1}{\omega_j}\mathbf{x}_j^T\mathbf{f}(\tau)\sin\left(\omega_j(t-\tau)\right)d\tau + B_j\cos\left(\omega_j t - \vartheta_j\right) \tag{3.91}$$

where

$$B_j = \sqrt{\left(\mathbf{x}_j^T\mathbf{M}\mathbf{q}_0\right)^2 + \left(\frac{\mathbf{x}_j^T\mathbf{M}\dot{\mathbf{q}}_0}{\omega_j}\right)^2} \tag{3.92}$$

$$\text{and} \quad \tan\vartheta_j = \frac{\mathbf{x}_j^T\mathbf{M}\dot{\mathbf{q}}_0}{\omega_j\left(\mathbf{x}_j^T\mathbf{M}\mathbf{q}_0\right)}. \tag{3.93}$$

Observe that the second part of Equation 3.91, i.e., the term $B_j\cos\left(\omega_j t - \vartheta_j\right)$ only depends on the initial conditions and independent of the applied loading.

3.4 PROPORTIONALLY DAMPED SYSTEMS

Equation of motion of a viscously damped system can be obtained from the Lagrange's equation and using the Rayleigh's dissipation function. The non-conservative forces can be obtained as

$$Q_{nc_k} = -\frac{\partial \mathbf{F}}{\partial \dot{q}_k}, \quad k = 1, \cdots, N \tag{3.94}$$

and consequently the equation of motion can be expressed as

$$\mathbf{M}\ddot{\mathbf{q}}(t) + \mathbf{C}\dot{\mathbf{q}}(t) + \mathbf{K}\mathbf{q}(t) = \mathbf{f}(t). \tag{3.95}$$

The aim is to solve this equation, together with the initial conditions, by classical modal analysis as described in Section 3.3.1. Using the modal transformation in (3.69), premultiplying Equation 3.95 by $\mathbf{X}^T$ and using the orthogonality relationships in (3.67) and (3.68), equation of motion of a damped system in the modal coordinates may be obtained as

$$\ddot{\mathbf{y}}(t) + \mathbf{X}^T \mathbf{C} \mathbf{X} \dot{\mathbf{y}}(t) + \mathbf{\Omega}^2 \mathbf{y}(t) = \tilde{\mathbf{f}}(t). \tag{3.96}$$

Clearly, unless $\mathbf{X}^T \mathbf{C} \mathbf{X}$ is a diagonal matrix, no advantage can be gained by employing modal analysis because the equations of motion will still be coupled. To solve this problem, it is common to assume *proportional damping* which we will discuss in some detail.

With the proportional damping assumption, the damping matrix $\mathbf{C}$ is simultaneously diagonalizable with $\mathbf{M}$ and $\mathbf{K}$. This implies that the damping matrix in the modal coordinate

$$\mathbf{C}' = \mathbf{X}^T \mathbf{C} \mathbf{X} \tag{3.97}$$

is a diagonal matrix. This matrix is also known as the modal damping matrix. The damping factors ζ_j are defined from the diagonal elements of the modal damping matrix as

$$C'_{jj} = 2\zeta_j \omega_j \quad \forall j = 1, \cdots, N. \tag{3.98}$$

Such a damping model, introduced by Lord Rayleigh [157], allows to analyze damped systems in very much the same manner as undamped systems since the equation of motion in the modal coordinates can be decoupled as

$$\ddot{y}_j(t) + 2\zeta_j \omega_j \dot{y}_j(t) + \omega_j^2 y_j(t) = \tilde{f}_j(t) \quad \forall j = 1, \cdots, N. \tag{3.99}$$

The proportional damping model expresses the damping matrix as a linear combination of the mass and stiffness matrices, that is

$$\mathbf{C} = \alpha_1 \mathbf{M} + \alpha_2 \mathbf{K} \tag{3.100}$$

where α_1 and α_2 are real scalars. This damping model is also known as "Rayleigh damping" or "classical damping." Modes of classically damped systems preserve the simplicity of the real normal modes as in the undamped case. Later, in a significant paper Caughey and O'Kelly [26] have shown that the classical damping can exist in more general situation.

3.4.1 CONDITION FOR PROPORTIONAL DAMPING

Caughey and O'Kelly [26] have derived the condition which the system matrices must satisfy so that viscously damped linear systems possess classical normal modes. Their result can be described in the following theorem

---

Theorem 3.1

A viscously damped system (3.95) possesses classical normal modes if and only if $\mathbf{C} \mathbf{M}^{-1} \mathbf{K} = \mathbf{K} \mathbf{M}^{-1} \mathbf{C}$. ∎

Outline of the Proof. See the original paper by Caughey and O'Kelly [26] for the detailed proof. Assuming $\mathbf{M}$ is not singular, premultiplying Equation 3.95 by $\mathbf{M}^{-1}$ we have

$$\mathbf{I}\ddot{\mathbf{q}}(t) + [\mathbf{M}^{-1}\mathbf{C}]\dot{\mathbf{q}}(t) + [\mathbf{M}^{-1}\mathbf{K}]\mathbf{q}(t) = \mathbf{M}^{-1}\mathbf{f}(t). \qquad (3.101)$$

For classical normal modes, 3.101 must be diagonalized by an orthogonal transformation. Two matrices $\mathbf{A}$ and $\mathbf{B}$ can be diagonalized by an orthogonal transformation if and only if the commute in product [16], i.e., $\mathbf{AB} = \mathbf{BA}$. Using this condition in 3.101 we have

$$[\mathbf{M}^{-1}\mathbf{C}][\mathbf{M}^{-1}\mathbf{K}] = [\mathbf{M}^{-1}\mathbf{K}][\mathbf{M}^{-1}\mathbf{C}],$$

$$\text{or} \quad \mathbf{CM}^{-1}\mathbf{K} = \mathbf{KM}^{-1}\mathbf{C} \quad \text{(premultiplying both sides by } \mathbf{M}\text{).} \qquad (3.102)$$

A modified and more general version of this theorem was proved in [1].

Example 3.1. Assume that a system's mass, stiffness and damping matrices are given by

$$\mathbf{M} = \begin{bmatrix} 1.0 & 1.0 & 1.0 \\ 1.0 & 2.0 & 2.0 \\ 1.0 & 2.0 & 3.0 \end{bmatrix}, \quad \mathbf{K} = \begin{bmatrix} 2 & -1 & 0.5 \\ -1 & 1.2 & 0.4 \\ 0.5 & 0.4 & 1.8 \end{bmatrix} \quad \text{and} \qquad (3.103)$$

$$\mathbf{C} = \begin{bmatrix} 15.25 & -9.8 & 3.4 \\ -9.8 & 6.48 & -1.84 \\ 3.4 & -1.84 & 2.22 \end{bmatrix}.$$

It may be verified that all the system matrices are positive definite. The mass-normalized undamped modal matrix is obtained as

$$\mathbf{X} = \begin{bmatrix} 0.4027 & -0.5221 & -1.2511 \\ 0.5845 & -0.4888 & 1.1914 \\ -0.1127 & 0.9036 & -0.4134 \end{bmatrix}. \qquad (3.104)$$

Since Caughey and O'Kelly's condition

$$\mathbf{KM}^{-1}\mathbf{C} = \mathbf{CM}^{-1}\mathbf{K} = \begin{bmatrix} 125.45 & -80.92 & 28.61 \\ -80.92 & 52.272 & -18.176 \\ 28.61 & -18.176 & 7.908 \end{bmatrix} \qquad (3.105)$$

is satisfied, the system possess classical normal modes and that $\mathbf{X}$ given in Equation 3.104 is the modal matrix. Because the system is positive definite, two other equivalent conditions, namely

$$\mathbf{MK}^{-1}\mathbf{C} = \mathbf{CK}^{-1}\mathbf{M} = \begin{bmatrix} 2.0 & -1.0 & 0.5 \\ -1.0 & 1.2 & 0.4 \\ 0.5 & 0.4 & 1.8 \end{bmatrix} \qquad (3.106)$$

and

$$\mathbf{MC}^{-1}\mathbf{K} = \mathbf{KC}^{-1}\mathbf{M} = \begin{bmatrix} 4.1 & 6.2 & 5.6 \\ 6.2 & 9.73 & 9.2 \\ 5.6 & 9.2 & 9.6 \end{bmatrix} \qquad (3.107)$$

are also satisfied.

3.4.2 GENERALIZED PROPORTIONAL DAMPING

Obtaining a damping matrix from "first principles" as with the mass and stiffness matrices is not possible for most systems. For this reason, assuming $\mathbf{M}$ and $\mathbf{K}$ are known, we often want to find $\mathbf{C}$ in terms of $\mathbf{M}$ and $\mathbf{K}$ such that the system still possesses classical normal modes. Of course, the earliest work along this line is the proportional damping shown in Equation 3.100 by Rayleigh [157]. It can be verified that, for positive definite systems, expressing $\mathbf{C}$ in such a way will always satisfy the condition given by 3.1. Caughey [25] proposed that a *sufficient* condition for the existence of classical normal modes is: if $\mathbf{M}^{-1}\mathbf{C}$ can be expressed in a series involving powers of $\mathbf{M}^{-1}\mathbf{K}$. His result generalized Rayleigh's result, which turns out to be the first two terms of the series. Later, Caughey and O'Kelly [26] proved that the series representation of damping

$$\mathbf{C} = \mathbf{M} \sum_{j=0}^{N-1} \alpha_j \left[\mathbf{M}^{-1}\mathbf{K}\right]^j \qquad (3.108)$$

is the *necessary and sufficient* condition for existence of classical normal modes for systems without any repeated roots. This series is now known as the "Caughey series."

A further generalized and useful form of proportional damping was proposed in [1]. It can be shown that we can express the damping matrix in the form

$$\mathbf{C} = \mathbf{M}\,\beta_1\left(\mathbf{M}^{-1}\mathbf{K}\right) + \mathbf{K}\,\beta_2\left(\mathbf{K}^{-1}\mathbf{M}\right) \qquad (3.109)$$

$$\text{or} \quad \mathbf{C} = \beta_3\left(\mathbf{K}\mathbf{M}^{-1}\right)\mathbf{M} + \beta_4\left(\mathbf{M}\mathbf{K}^{-1}\right)\mathbf{K} \qquad (3.110)$$

The functions $\beta_i(\bullet)$, $i = 1, \cdots, 4$ can have very general forms—they may consist of an arbitrary number of multiplications, divisions, summations, subtractions or powers of any other functions or can even be functional compositions. Thus, any conceivable form of analytic functions that are valid for scalars can be used in Equations (3.109) and (3.110). In a natural way, common restrictions applicable to scalar functions are also valid, for example logarithm of a negative number is not permitted. Although the functions $\beta_i(\bullet)$, $i = 1, \cdots, 4$ are general, the expression of $\mathbf{C}$ in 3.109 or 3.110 gets restricted because of the special nature of the *arguments* in the functions. As a consequence, $\mathbf{C}$ represented in 3.109 or 3.110 does not cover the whole $\mathbb{R}^{N \times N}$, which is well known that many damped systems do not possess classical normal modes.

Rayleigh's result (3.100) can be obtained directly from Equation 3.109 or (3.110) as a very special case by choosing each matrix function $\beta_i(\bullet)$ as real scalar times an identity matrix, that is

$$\beta_i(\bullet) = \alpha_i \mathbf{I}. \qquad (3.111)$$

The damping matrix expressed in Equation 3.109 or 3.110 provides a new way of interpreting the "Rayleigh damping" or "proportional damping" where the identity matrices (always) associated in the right or left side of $\mathbf{M}$ and $\mathbf{K}$ are replaced by arbitrary matrix functions $\beta_i(\bullet)$ with proper arguments. This kind of damping model will be called *generalized proportional damping*. We call the representation in Equation 3.109 *right-functional form* and that in Equation 3.110 *left-functional form*. Caughey series 3.108 is an example of right functional form. Note that if $\mathbf{M}$ or $\mathbf{K}$ is singular then the argument involving its corresponding inverse has to be removed from the functions. We will call the functions $\beta_i(\bullet)$ as *proportional damping functions* which are consistent with the definition of proportional damping constants (α_i) in Rayleigh's model. From this discussion we have the following general result for damped linear systems

---

Theorem 3.2

A viscously damped positive definite linear system possesses classical normal modes if and only if $\mathbf{C}$ can be represented by

(a) $\mathbf{C} = \mathbf{M}\,\beta_1\left(\mathbf{M}^{-1}\mathbf{K}\right) + \mathbf{K}\,\beta_2\left(\mathbf{K}^{-1}\mathbf{M}\right)$, or
(b) $\mathbf{C} = \beta_3\left(\mathbf{K}\mathbf{M}^{-1}\right)\mathbf{M} + \beta_4\left(\mathbf{M}\mathbf{K}^{-1}\right)\mathbf{K}$
    for any $\beta_i(\bullet), i = 1, \cdots, 4$.

∎

Proof. Consider the 'if' part first. Suppose $\mathbf{X}$ is the mass normalized modal matrix and $\mathbf{\Omega}$ is the diagonal matrix containing the undamped natural frequencies. By the definitions of these quantities we have

$$\mathbf{X}^T\mathbf{M}\mathbf{X} = \mathbf{I} \tag{3.112}$$

$$\text{and} \quad \mathbf{X}^T\mathbf{K}\mathbf{X} = \mathbf{\Omega}^2. \tag{3.113}$$

From these equations one obtains

$$\mathbf{M} = \mathbf{X}^{-T}\mathbf{X}^{-1}, \quad \mathbf{K} = \mathbf{X}^{-T}\mathbf{\Omega}^2\mathbf{X}^{-1} \tag{3.114}$$

$$\mathbf{M}^{-1}\mathbf{K} = \mathbf{X}\mathbf{\Omega}^2\mathbf{X}^{-1} \quad \text{and} \quad \mathbf{K}^{-1}\mathbf{M} = \mathbf{X}\mathbf{\Omega}^{-2}\mathbf{X}^{-1}. \tag{3.115}$$

Because the functions $\beta_1(\bullet)$ and $\beta_2(\bullet)$ are assumed to be analytic in the neighborhood of all the eigenvalues of $\mathbf{M}^{-1}\mathbf{K}$ and $\mathbf{K}^{-1}\mathbf{M}$ respectively, they can be expressed in polynomial forms using the Taylor series expansion. Following Bellman [16] (Chapter 6) we may obtain

$$\beta_1\left(\mathbf{M}^{-1}\mathbf{K}\right) = \mathbf{X}\,\beta_1\left(\mathbf{\Omega}^2\right)\mathbf{X}^{-1} \tag{3.116}$$

$$\text{and} \quad \beta_2\left(\mathbf{K}^{-1}\mathbf{M}\right) = \mathbf{X}\,\beta_2\left(\mathbf{\Omega}^{-2}\right)\mathbf{X}^{-1}. \tag{3.117}$$

A viscously damped system will possess classical normal modes if $\mathbf{X}^T\mathbf{C}\mathbf{X}$ is a diagonal matrix. Considering expression (a) in the theorem and using

Equations 3.114 and 3.115 we have

$$\mathbf{X}^T\mathbf{C}\mathbf{X} = \mathbf{X}^T\left[\mathbf{M}\,\beta_1\left(\mathbf{M}^{-1}\mathbf{K}\right) + \mathbf{K}\,\beta_2\left(\mathbf{K}^{-1}\mathbf{M}\right)\right]\mathbf{X}$$
$$= \mathbf{X}^T\left[\mathbf{X}^{-T}\mathbf{X}^{-1}\,\beta_1\left(\mathbf{M}^{-1}\mathbf{K}\right) + \mathbf{X}^{-T}\mathbf{\Omega}^2\mathbf{X}^{-1}\,\beta_2\left(\mathbf{K}^{-1}\mathbf{M}\right)\right]\mathbf{X}.$$
$$(3.118)$$

Utilizing (3.116) and (3.117) and carrying out the matrix multiplications, Equation 3.118 reduces to

$$\mathbf{X}^T\mathbf{C}\mathbf{X} = \left[\mathbf{X}^{-1}\,\mathbf{X}\,\beta_1\left(\mathbf{\Omega}^2\right)\mathbf{X}^{-1} + \mathbf{\Omega}^2\mathbf{X}^{-1}\,\mathbf{X}\,\beta_2\left(\mathbf{\Omega}^{-2}\right)\mathbf{X}^{-1}\right]\mathbf{X}$$
$$= \beta_1\left(\mathbf{\Omega}^2\right) + \mathbf{\Omega}^2\beta_2\left(\mathbf{\Omega}^{-2}\right).$$
$$(3.119)$$

Equation 3.119 clearly shows that $\mathbf{X}^T\mathbf{C}\mathbf{X}$ is a diagonal matrix.

To prove the the 'only if' part, suppose

$$\mathbf{P} = \mathbf{X}^T\mathbf{C}\mathbf{X} \qquad (3.120)$$

is a general matrix (not necessary diagonal). Then there exist a non-zero matrix $\mathbf{S}$ such that (similarity transform)

$$\mathbf{S}^{-1}\mathbf{P}\mathbf{S} = \mathcal{D} \qquad (3.121)$$

where $\mathcal{D}$ is a diagonal matrix. Using (3.119) and (3.120) we have

$$\mathbf{S}^{-1}\mathcal{D}_1\mathbf{S} = \mathcal{D} \qquad (3.122)$$

where $\mathcal{D}_1$ is another diagonal matrix. Equation 3.122 indicates that two diagonal matrices are related by a similarity transformation. This can only happen when they are the same and the transformation matrix is an identity matrix, that is $\mathbf{S} = \mathbf{I}$. Using this in Equation 3.121 proves that $\mathbf{P}$ must be a diagonal matrix. $\qquad\qquad\qquad\qquad\qquad\qquad\qquad\qquad\qquad\qquad\qquad\qquad\square$

Example 3.2. To investigate the applicability of this result, it will be shown that the linear dynamic system satisfying the following equations of free vibration

$$\mathbf{M}\ddot{\mathbf{q}} + \left[\mathbf{M}e^{-\left(\mathbf{M}^{-1}\mathbf{K}\right)^2/2}\sinh(\mathbf{K}^{-1}\mathbf{M}\ln(\mathbf{M}^{-1}\mathbf{K})^{2/3})\right.$$
$$\left. + \mathbf{K}\cos^2(\mathbf{K}^{-1}\mathbf{M})\sqrt[4]{\mathbf{K}^{-1}\mathbf{M}}\,\tan^{-1}\frac{\sqrt{\mathbf{M}^{-1}\mathbf{K}}}{\pi}\right]\dot{\mathbf{q}} + \mathbf{K}\mathbf{q} = 0$$
$$(3.123)$$

possesses classical normal modes and can be analysed using modal analysis. Here $\mathbf{M}$ and $\mathbf{K}$ are the same as example 3.1.

Direct calculation shows

$$\mathbf{C} = \begin{bmatrix} -67.9188 & -104.8208 & -95.9566 \\ -104.8208 & -161.1897 & -147.7378 \\ -95.9566 & -147.7378 & -135.2643 \end{bmatrix}. \qquad (3.124)$$

Using the modal matrix calculated before in (3.104), we obtain

$$\mathbf{X}^T \mathbf{CX} = \begin{bmatrix} -88.9682 & 0.0 & 0.0 \\ 0.0 & 0.0748 & 0.0 \\ 0.0 & 0.0 & 0.5293 \end{bmatrix}, \tag{3.125}$$

which is a diagonal matrix. Analytically the modal damping factors can be obtained as

$$2\zeta_j \omega_j = e^{-\omega_j^4/2} \sinh\left(\frac{1}{\omega_j^2} \ln \frac{4}{3} \omega_j \right) + \omega_j^2 \cos^2\left(\frac{1}{\omega_j^2} \right) \frac{1}{\sqrt{\omega_j}} \tan^{-1} \frac{\omega_j}{\pi}. \tag{3.126}$$

A natural question which arises in the context of the generalized proportional damping is how to obtain the damping functions from experimental modal analysis. This will be addressed later in this book.

3.4.3 DYNAMIC RESPONSE

Dynamic response of proportionally damped systems can be obtained in a similar way to that of undamped systems discussed before in 3.3.2. Both frequency-domain and and time-domain approaches are shown here.

3.4.3.1 Frequency-Domain Analysis

Taking the Laplace transform of (3.95) and considering the initial conditions in (3.54) one has

$$s^2 \mathbf{M}\bar{\mathbf{q}} - s\mathbf{M}\mathbf{q}_0 - \mathbf{M}\dot{\mathbf{q}}_0 + s\mathbf{C}\bar{\mathbf{q}} - \mathbf{C}\mathbf{q}_0 + \mathbf{K}\bar{\mathbf{q}} = \bar{\mathbf{f}}(s) \tag{3.127}$$

$$\text{or} \quad \left[s^2 \mathbf{M} + s\mathbf{C} + \mathbf{K} \right] \bar{\mathbf{q}} = \bar{\mathbf{f}}(s) + \mathbf{M}\dot{\mathbf{q}}_0 + \mathbf{C}\mathbf{q}_0 + s\mathbf{M}\mathbf{q}_0. \tag{3.128}$$

Consider the modal damping matrix

$$\mathbf{C}' = \mathbf{X}^T \mathbf{CX} = 2\zeta\mathbf{\Omega} \tag{3.129}$$

where

$$\zeta = \text{diag}\left[\zeta_1, \zeta_2, \cdots, \zeta_N \right] \in \mathbb{R}^{N \times N} \tag{3.130}$$

is the diagonal matrix containing the modal damping factors. Using the mode orthogonality relationships and following the procedure similar to undamped systems, it is easy to show that

$$\bar{\mathbf{q}}(s) = \mathbf{X} \left[s^2 \mathbf{I} + 2s\zeta\mathbf{\Omega} + \mathbf{\Omega}^2 \right]^{-1} \mathbf{X}^T \left\{ \bar{\mathbf{f}}(s) + \mathbf{M}\dot{\mathbf{q}}_0 + \mathbf{C}\mathbf{q}_0 + s\mathbf{M}\mathbf{q}_0 \right\}. \tag{3.131}$$

The dynamic response in the frequency domain can be obtained by substituting $s = i\omega$. Note that $\left[s^2 \mathbf{I} + 2s\zeta\mathbf{\Omega} + \mathbf{\Omega}^2 \right]$ is a diagonal matrix and therefore its inverse is easy to obtain. Following the procedure similar to undamped systems, the transfer function matrix can be obtained as

$$\mathbf{H}(i\omega) = \mathbf{X}\left[-\omega^2\mathbf{I} + 2i\omega\zeta\mathbf{\Omega} + \mathbf{\Omega}^2\right]^{-1}\mathbf{X}^T = \sum_{j=1}^{N}\frac{\mathbf{x}_j\mathbf{x}_j^T}{-\omega^2 + 2i\omega\zeta_j\omega_j + \omega_j^2}.$$

(3.132)

Using this, the dynamic response in the frequency domain can be conveniently represented from Equation 3.131 as

$$\bar{\mathbf{q}}(i\omega) = \sum_{j=1}^{N}\frac{\mathbf{x}_j^T\bar{\mathbf{f}}(i\omega) + \mathbf{x}_j^T\mathbf{M}\dot{\mathbf{q}}_0 + \mathbf{x}_j^T\mathbf{C}\mathbf{q}_0 + i\omega\mathbf{x}_j^T\mathbf{M}\mathbf{q}_0}{-\omega^2 + 2i\omega\zeta_j\omega_j + \omega_j^2}\mathbf{x}_j.$$

(3.133)

Therefore, like undamped systems, the dynamic response of proportionally damped system can also be expressed as a linear combination of the undamped mode shapes. This shows how classical modal analysis can be used for proportionally damped systems.

3.4.3.2 Time-Domain Analysis

Rewrite the frequency-domain response in (3.133) in the Laplace domain as

$$\bar{\mathbf{q}}(s) = \sum_{j=1}^{N}\Big(\frac{\mathbf{x}_j^T\bar{\mathbf{f}}(s)}{s^2 + 2s\zeta_j\omega_j + \omega_j^2} + \frac{\mathbf{x}_j^T\mathbf{M}\dot{\mathbf{q}}_0 + \mathbf{x}_j^T\mathbf{C}\mathbf{q}_0}{s^2 + 2s\zeta_j\omega_j + \omega_j^2} + \quad\quad (3.134)$$

$$\frac{s}{s^2 + 2s\zeta_j\omega_j + \omega_j^2}\mathbf{x}_j^T\mathbf{M}\mathbf{q}_0\Big)\mathbf{x}_j.$$

We can reorganize the denominator as

$$s^2 + 2s\zeta_j\omega_j + \omega_j^2 = (s + \zeta_j\omega_j)^2 - (\zeta_j\omega_j)^2 + \omega_j^2 = (s + \zeta_j\omega_j)^2 + \omega_{d_j}^2 \quad (3.135)$$

where the damped natural frequency for the j-th mode is given by

$$\omega_{d_j} = \omega_j\sqrt{1 - \zeta_j^2} \quad\quad (3.136)$$

Now we use the expressions of the inverse Laplace transforms in Equations 3.39 and 3.40 with α and β standing for $\zeta\omega_j$ and ω_{d_j} respectively. Taking the inverse Laplace transform of the dynamic response (3.134) we have

$$\mathbf{q}(t) = \mathcal{L}^{-1}\left[\bar{\mathbf{q}}(s)\right] = \sum_{j=1}^{N}a_j(t)\mathbf{x}_j \quad\quad (3.137)$$

where the time dependent constants are given by

$$
a_j(t) = \mathcal{L}^{-1} \left[\frac{\mathbf{x}_j^T \bar{\mathbf{f}}(s)}{(s + \zeta_j \omega_j)^2 + \omega_{d_j}^2} \right]
$$

$$
+ \mathcal{L}^{-1} \left[\frac{1}{(s + \zeta_j \omega_j)^2 + \omega_{d_j}^2} \right] \left(\mathbf{x}_j^T \mathbf{M} \dot{\mathbf{q}}_0 + \mathbf{x}_j^T \mathbf{C} \mathbf{q}_0 \right)
$$

$$
+ \mathcal{L}^{-1} \left[\frac{s}{(s + \zeta_j \omega_j)^2 + \omega_{d_j}^2} \right] \mathbf{x}_j^T \mathbf{M} \mathbf{q}_0
$$

$$
= \int_0^t \frac{1}{\omega_{d_j}} \mathbf{x}_j^T \mathbf{f}(\tau) e^{-\zeta_j \omega_j (t - \tau)} \sin \left(\omega_{d_j} (t - \tau) \right) d\tau
$$

$$
+ \frac{e^{-\zeta_j \omega_j t}}{\omega_{d_j}} \sin(\omega_{d_j} t) \left(\mathbf{x}_j^T \mathbf{M} \dot{\mathbf{q}}_0 + \mathbf{x}_j^T \mathbf{C} \mathbf{q}_0 \right)
$$

$$
+ \left\{ e^{-\zeta_j \omega_j t} \cos \left(\omega_{d_j} t \right) - \frac{\zeta_j \omega_j \, e^{-\zeta_j \omega_j t} \sin \left(\omega_{d_j} t \right)}{\omega_{d_j}} \right\} \mathbf{x}_j^T \mathbf{M} \mathbf{q}_0
$$

$$
= \int_0^t \frac{1}{\omega_{d_j}} \mathbf{x}_j^T \mathbf{f}(\tau) e^{-\zeta_j \omega_j (t - \tau)} \sin \left(\omega_{d_j} (t - \tau) \right) d\tau
$$

$$
+ e^{-\zeta_j \omega_j t}
$$

$$
\left(\frac{\sin(\omega_{d_j} t)}{\omega_{d_j}} \left(\mathbf{x}_j^T \mathbf{M} \dot{\mathbf{q}}_0 + \mathbf{x}_j^T \mathbf{C} \mathbf{q}_0 - \zeta_j \omega_j \mathbf{x}_j^T \mathbf{M} \mathbf{q}_0 \right) + \cos \left(\omega_{d_j} t \right) \left(\mathbf{x}_j^T \mathbf{M} \mathbf{q}_0 \right) \right)
$$

$$
(3.138)
$$

The convolution theorem is used to obtain the inverse Laplace transform of the first part. The formulae shown in (3.39) and (3.40) are used to obtain the inverse Laplace transform of the second and third parts. The coefficients associated with $\sin(\omega_{d_j} t)$ in the preceding equation can be simplified. Suppose the initial displacement in the modal coordinate is $\mathbf{y}_0$ so that $\mathbf{q}_0 = \mathbf{X} \mathbf{y}_0$. Therefore

$$
\mathbf{x}_j^T \mathbf{M} \mathbf{q}_0 = \mathbf{x}_j^T \mathbf{M} \mathbf{X} \mathbf{y}_0 = \mathbf{x}_j^T \mathbf{M} \left[\mathbf{x}_1, \mathbf{x}_2, \cdots, \mathbf{x}_N \right] \mathbf{y}_0 = y_{0_j} \qquad (3.139)
$$

which is the initial displacement of the jth modal coordinate. We have used the mass orthonormality properties of the eigenvectors in (3.63). Using this and recalling the modal damping matrix in (3.129) we have

$$
\mathbf{x}_j^T \mathbf{C} \mathbf{q}_0 - \zeta_j \omega_j \mathbf{x}_j^T \mathbf{M} \mathbf{q}_0 = \qquad\qquad\qquad (3.140)
$$
$$
\mathbf{x}_j^T \mathbf{C} \mathbf{X} \mathbf{y}_0 - \zeta_j \omega_j \mathbf{x}_j^T \mathbf{M} \mathbf{X} \mathbf{y}_0 = 2 \zeta_j \omega_j y_{0_j} - \zeta_j \omega_j y_{0_j} = \zeta_j \omega_j y_{0_j}
$$

Using the Equation 3.139 we can establish

$$
\mathbf{x}_j^T \mathbf{C} \mathbf{q}_0 - \zeta_j \omega_j \mathbf{x}_j^T \mathbf{M} \mathbf{q}_0 = \zeta_j \omega_j \mathbf{x}_j^T \mathbf{M} \mathbf{q}_0 \quad \text{or} \quad \mathbf{x}_j^T \mathbf{C} \mathbf{q}_0 = 2 \zeta_j \omega_j \mathbf{x}_j^T \mathbf{M} \mathbf{q}_0 \quad (3.141)
$$

Substituting this simplification in (3.138) and collecting the terms associated with $\sin\left(\omega_{d_j} t\right)$ and $\cos\left(\omega_{d_j} t\right)$, the resulting expression can be simplified as

$$a_j(t) = \int_0^t \frac{1}{\omega_{d_j}} \mathbf{x}_j^T \mathbf{f}(\tau) e^{-\zeta_j \omega_j (t-\tau)}$$

(3.142)

$$\sin\left(\omega_{d_j}(t-\tau)\right) d\tau + e^{-\zeta_j \omega_j t} B_j \cos\left(\omega_{d_j} t - \vartheta_j\right)$$

where

$$B_j = \sqrt{\left(\mathbf{x}_j^T \mathbf{M} \mathbf{q}_0\right)^2 + \left(\frac{\mathbf{x}_j^T \mathbf{M} \dot{\mathbf{q}}_0 + \frac{1}{2}\mathbf{x}_j^T \mathbf{C} \mathbf{q}_0}{\omega_{d_j}}\right)^2}$$

(3.143)

$$\text{and} \quad \tan \vartheta_j = \frac{\mathbf{x}_j^T \mathbf{M} \dot{\mathbf{q}}_0 + \frac{1}{2}\mathbf{x}_j^T \mathbf{C} \mathbf{q}_0}{\omega_{d_j}\left(\mathbf{x}_j^T \mathbf{M} \mathbf{q}_0\right)}$$

(3.144)

All the terms appearing in the preceding equations can be obtained from the natural frequencies, modal damping factors, forcing function and initial conditions. The modal amplitudes and phase angles in (3.143) and (3.144) can also be expressed in terms of the initial conditions in the modal coordinates as

$$B_j = \sqrt{y_{0_j}^2 + \left(\frac{\dot{y}_{0_j} + \zeta_j \omega_j y_{0_j}}{\omega_{d_j}}\right)^2} \quad \text{and} \quad \tan \vartheta_j = \frac{\dot{y}_{0_j} + \zeta_j \omega_j y_{0_j}}{\omega_{d_j} y_{0_j}}.$$

(3.145)

This can be computationally efficient as $\mathbf{y}_0$ and $\dot{\mathbf{y}}_0$ can be stored a priori. In practice often only first few modes are used in the modal superposition expression in (3.137). In the special case when there is no damping, by substituting $\mathbf{C} = \mathbf{O}$ it can be seen that the results derived here reduces to the ones derived in Section 3.3.2.2.

Example 3.3. Figure 3.4 shows a 3-DOF spring-mass system. The mass of each block is m Kg and the stiffness of each spring is k N/m. The viscous

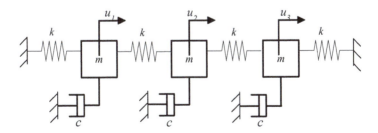

Figure 3.4 3-DOF damped spring-mass system with dampers attached to the ground.

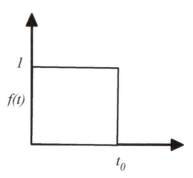

Figure 3.5   Unit step forcing, $t_0 = \frac{2\pi}{\omega_1}$.

damping constant of the damper associated with each block is c Ns/m. The aim is to obtain the dynamic response for the following load cases:

1. When only the first mass (DOF 1) is subjected to an unit step input (see 3.5) so that $\mathbf{f}(t) = \{f(t), 0, 0\}^T$ and $f(t) = 1 - U(t - t_0)$ with $t_0 = \frac{2\pi}{\omega_1}$ where ω_1 is the first undamped natural frequency of the system and $U(\bullet)$ is the unit step function.
2. When only the second mass (DOF 2) is subjected to unit initial displacement, i.e., $\mathbf{q}_0 = \{0, 1, 0\}^T$.
3. When only the second and the third masses (DOF 3) are subjected to unit initial velocities, i.e., $\dot{\mathbf{q}}_0 = \{0, 1, 1\}^T$.
4. When all three of the above loading are acting together on the system.

This problem can be solved by using the following steps

   I. *Obtain the System Matrices:* The mass, stiffness and the damping matrices are given by

$$\mathbf{M} = \begin{bmatrix} m & 0 & 0 \\ 0 & m & 0 \\ 0 & 0 & m \end{bmatrix}, \quad \mathbf{K} = \begin{bmatrix} 2k & -k & 0 \\ -k & 2k & -k \\ 0 & -k & 2k \end{bmatrix} \quad \text{and} \quad (3.146)$$

$$\mathbf{C} = \begin{bmatrix} c & 0 & 0 \\ 0 & c & 0 \\ 0 & 0 & c \end{bmatrix}. \quad (3.147)$$

Note that the damping matrix is mass proportional, so that the system is proportionally damped.

   II. *Obtain the undamped natural frequencies:* For notational convenience assume that the eigenvalues $\lambda_j = \omega_j^2$. The 3-DOF system has three eigenvalues and they are the roots of the following characteristic equation

$$\det \left[\mathbf{K} - \lambda \mathbf{M} \right] = 0. \quad (3.148)$$

Using the mass and the stiffness matrices from (3.146), this can be simplified as

$$\det \begin{bmatrix} 2k - \lambda m & -k & 0 \\ -k & 2k - \lambda m & -k \\ 0 & -k & 2k - \lambda m \end{bmatrix} = 0$$

$$\text{or} \quad m \det \begin{bmatrix} 2\alpha - \lambda & -\alpha & 0 \\ -\alpha & 2\alpha - \lambda & -\alpha \\ 0 & -\alpha & 2\alpha - \lambda \end{bmatrix} = 0 \quad \text{where} \quad \alpha = \frac{k}{m}.$$

(3.149)

Expanding the determinant in (3.149) we have

$$(2\alpha - \lambda)\left\{ (2\alpha - \lambda)^2 - \alpha^2 \right\} - \alpha\alpha(2\alpha - \lambda) = 0$$

$$\text{or} \quad (2\alpha - \lambda)\left\{ (2\alpha - \lambda)^2 - 2\alpha^2 \right\} = 0$$

$$\text{or} \quad (2\alpha - \lambda)\left\{ (2\alpha - \lambda)^2 - \left(\sqrt{2}\alpha\right)^2 \right\} = 0 \qquad (3.150)$$

$$\text{or} \quad (2\alpha - \lambda)\left(2\alpha - \lambda - \sqrt{2}\alpha \right)\left(2\alpha - \lambda + \sqrt{2}\alpha \right) = 0$$

$$\text{or} \quad (2\alpha - \lambda)\left(\left(2 - \sqrt{2}\right)\alpha - \lambda \right)\left(\left(2 + \sqrt{2}\right)\alpha - \lambda \right) = 0.$$

It implies that the three roots (in the increasing order) are

$$\lambda_1 = \left(2 - \sqrt{2}\right)\alpha, \quad \lambda_2 = 2\alpha, \quad \text{and} \quad \lambda_3 = \left(2 + \sqrt{2}\right)\alpha. \qquad (3.151)$$

Since $\lambda_j = \omega_j^2$, the natural frequencies are

$$\omega_1 = \sqrt{\left(2 - \sqrt{2}\right)\alpha}, \quad \omega_2 = \sqrt{2\alpha}, \quad \text{and} \quad \omega_3 = \sqrt{\left(2 + \sqrt{2}\right)\alpha}. \quad (3.152)$$

III. *Obtain the undamped mode shapes:* From (3.55) the eigenvalue equation can be written as

$$[\mathbf{K} - \lambda_j \mathbf{M}]\,\mathbf{x}_j = 0. \qquad (3.153)$$

For this problem, substituting $\mathbf{K}$ and $\mathbf{M}$ from (3.146) and dividing by m we have

$$\begin{bmatrix} 2\alpha - \lambda & -\alpha & 0 \\ -\alpha & 2\alpha - \lambda & -\alpha \\ 0 & -\alpha & 2\alpha - \lambda \end{bmatrix} \begin{Bmatrix} x_{1j} \\ x_{2j} \\ x_{3j} \end{Bmatrix} = 0. \qquad (3.154)$$

Here x_{1j}, x_{2j} and x_{3j} are the three components of jth eigenvector corresponding to the three masses. To obtain $\mathbf{x}_j$ we need to substitute λ_j from Equation (3.151) in the above equation and solve for each component of $\mathbf{x}_j$ for every j.

The first eigenvector, $j = 1$:
Substituting $\lambda = \lambda_1 = \left(2 - \sqrt{2}\right)\alpha$ in (3.154) we have

$$
\begin{bmatrix}
2\alpha - \left(2 - \sqrt{2}\right)\alpha & -\alpha & 0 \\
-\alpha & 2\alpha - \left(2 - \sqrt{2}\right)\alpha & -\alpha \\
0 & -\alpha & 2\alpha - \left(2 - \sqrt{2}\right)\alpha
\end{bmatrix}
\begin{Bmatrix}
x_{11} \\
x_{21} \\
x_{31}
\end{Bmatrix} = 0
$$

(3.155)

or
$$
\begin{bmatrix}
\sqrt{2}\alpha & -\alpha & 0 \\
-\alpha & \sqrt{2}\alpha & -\alpha \\
0 & -\alpha & \sqrt{2}\alpha
\end{bmatrix}
\begin{Bmatrix}
x_{11} \\
x_{21} \\
x_{31}
\end{Bmatrix} = 0
\quad \text{or} \quad
\begin{bmatrix}
\sqrt{2} & -1 & 0 \\
-1 & \sqrt{2} & -1 \\
0 & -1 & \sqrt{2}
\end{bmatrix}
\begin{Bmatrix}
x_{11} \\
x_{21} \\
x_{31}
\end{Bmatrix} = 0.
$$

(3.156)

This can be separated into three equations as

$$
\sqrt{2}x_{11} - x_{21} = 0, \quad -x_{11} + \sqrt{2}x_{21} - x_{31} = 0 \quad \text{and} \quad -x_{21} + \sqrt{2}x_{31} = 0.
$$

(3.157)

These three equations cannot be solved uniquely but once we fix one element, the other two elements can be expressed in terms of it. This implies that the ratios between the modal amplitudes are unique. Solving the system of linear equations (3.157) we have $x_{21} = \sqrt{2}x_{11}$ and $x_{21} = \sqrt{2}x_{31}$, that is $x_{11} = x_{31} = \gamma_1$ (say). Therefore the first eigenvector is given by

$$
\mathbf{x}_1 = \begin{Bmatrix} x_{11} \\ x_{21} \\ x_{31} \end{Bmatrix} = \gamma_1 \begin{Bmatrix} 1 \\ \sqrt{2} \\ 1 \end{Bmatrix}.
$$

(3.158)

The constant γ_1 can be obtained from the mass normalization condition in (3.63), that is

$$
\mathbf{x}_1^T \mathbf{M} \mathbf{x}_1 = 1 \quad \text{or} \quad \gamma_1 \begin{Bmatrix} 1 \\ \sqrt{2} \\ 1 \end{Bmatrix}^T \begin{bmatrix} m & 0 & 0 \\ 0 & m & 0 \\ 0 & 0 & m \end{bmatrix} \gamma_1 \begin{Bmatrix} 1 \\ \sqrt{2} \\ 1 \end{Bmatrix} = 1
$$

(3.159)

$$
\text{or} \quad \gamma_1^2 m \begin{Bmatrix} 1 \\ \sqrt{2} \\ 1 \end{Bmatrix}^T \begin{bmatrix} 1 & 0 & 0 \\ 0 & 1 & 0 \\ 0 & 0 & 1 \end{bmatrix} \begin{Bmatrix} 1 \\ \sqrt{2} \\ 1 \end{Bmatrix} = 1
$$

(3.160)

$$
\text{or} \quad \gamma_1^2 m \left(1 + \sqrt{2}\sqrt{2} + 1\right) = 1 \quad \text{that is} \quad \gamma_1 = \frac{1}{2\sqrt{m}}.
$$

(3.161)

Thus the mass normalized first eigenvector is given by

$$
\mathbf{x}_1 = \frac{1}{2\sqrt{m}} \begin{Bmatrix} 1 \\ \sqrt{2} \\ 1 \end{Bmatrix}.
$$

(3.162)

The second eigenvector, $j = 2$:

Substituting $\lambda = \lambda_2 = 2\alpha$ in (3.154) we have

$$\begin{bmatrix} 2\alpha - 2\alpha & -\alpha & 0 \\ -\alpha & 2\alpha - 2\alpha & -\alpha \\ 0 & -\alpha & 2\alpha - 2\alpha \end{bmatrix} \begin{Bmatrix} x_{12} \\ x_{22} \\ x_{32} \end{Bmatrix} = 0 \quad (3.163)$$

or $\quad \begin{bmatrix} 0 & -\alpha & 0 \\ -\alpha & 0 & -\alpha \\ 0 & -\alpha & 0 \end{bmatrix} \begin{Bmatrix} x_{12} \\ x_{22} \\ x_{32} \end{Bmatrix} = 0 \quad$ or $\quad \begin{bmatrix} 0 & 1 & 0 \\ 1 & 0 & 1 \\ 0 & 1 & 0 \end{bmatrix} \begin{Bmatrix} x_{12} \\ x_{22} \\ x_{32} \end{Bmatrix} = 0. \quad (3.164)$

This implies that $x_{22} = 0$ and $x_{12} = -x_{32} = \gamma_2$ (say). Therefore the second eigenvector is given by

$$\mathbf{x}_2 = \begin{Bmatrix} x_{12} \\ x_{22} \\ x_{32} \end{Bmatrix} = \gamma_2 \begin{Bmatrix} 1 \\ 0 \\ -1 \end{Bmatrix}. \quad (3.165)$$

Using the mass normalization condition

$$\mathbf{x}_2^T \mathbf{M} \mathbf{x}_2 = 1 \quad \text{or} \quad \gamma_2 \begin{Bmatrix} 1 \\ 0 \\ -1 \end{Bmatrix}^T \begin{bmatrix} m & 0 & 0 \\ 0 & m & 0 \\ 0 & 0 & m \end{bmatrix} \gamma_2 \begin{Bmatrix} 1 \\ 0 \\ -1 \end{Bmatrix} = 1 \quad (3.166)$$

or $\quad \gamma_2^2 m (1 + 1) = 1 \quad$ that is $\quad \gamma_2 = \dfrac{1}{\sqrt{2m}} = \dfrac{\sqrt{2}}{2\sqrt{m}}. \quad (3.167)$

Thus, the mass normalized second eigenvector is given by

$$\mathbf{x}_2 = \frac{1}{2\sqrt{m}} \begin{Bmatrix} \sqrt{2} \\ 0 \\ -\sqrt{2} \end{Bmatrix}. \quad (3.168)$$

The third eigenvector, $j = 3$:

Substituting $\lambda = \lambda_3 = \left(2 + \sqrt{2}\right)\alpha$ in (3.154) we have

$$\begin{bmatrix} 2\alpha - \left(2 + \sqrt{2}\right)\alpha & -\alpha & 0 \\ -\alpha & 2\alpha - \left(2 + \sqrt{2}\right)\alpha & -\alpha \\ 0 & -\alpha & 2\alpha - \left(2 + \sqrt{2}\right)\alpha \end{bmatrix} \begin{Bmatrix} x_{13} \\ x_{23} \\ x_{33} \end{Bmatrix} = 0$$

$$(3.169)$$

$$\begin{bmatrix} -\sqrt{2}\alpha & -\alpha & 0 \\ -\alpha & -\sqrt{2}\alpha & -\alpha \\ 0 & -\alpha & -\sqrt{2}\alpha \end{bmatrix} \begin{Bmatrix} x_{13} \\ x_{23} \\ x_{33} \end{Bmatrix} = 0$$

$$\begin{bmatrix} \sqrt{2} & 1 & 0 \\ 1 & \sqrt{2} & 1 \\ 0 & 1 & \sqrt{2} \end{bmatrix} \begin{Bmatrix} x_{13} \\ x_{23} \\ x_{33} \end{Bmatrix} = 0.$$

This implies that

$$\sqrt{2}x_{13} + x_{23} = 0 \Rightarrow x_{23} = -\sqrt{2}x_{13}$$

$$x_{13} + \sqrt{2}x_{23} + x_{33} = 0 \tag{3.170}$$

$$\text{and} \quad x_{23} + \sqrt{2}x_{33} = 0 \Rightarrow x_{23} = -\sqrt{2}x_{33}$$

ie., $x_{13} = x_{33} = \gamma_3$ (say). Therefore the third eigenvector is given by

$$\mathbf{x}_3 = \begin{Bmatrix} x_{13} \\ x_{23} \\ x_{33} \end{Bmatrix} = \gamma_3 \begin{Bmatrix} 1 \\ -\sqrt{2} \\ 1 \end{Bmatrix}. \tag{3.171}$$

Using the mass normalization condition

$$\mathbf{x}_3^T \mathbf{M} \mathbf{x}_3 = 1 \quad \text{or} \quad \gamma_3 \begin{Bmatrix} 1 \\ -\sqrt{2} \\ 1 \end{Bmatrix}^T \begin{bmatrix} m & 0 & 0 \\ 0 & m & 0 \\ 0 & 0 & m \end{bmatrix} \gamma_3 \begin{Bmatrix} 1 \\ -\sqrt{2} \\ 1 \end{Bmatrix} = 1 \tag{3.172}$$

$$\text{or} \quad \gamma_3^2 m \begin{Bmatrix} 1 \\ -\sqrt{2} \\ 1 \end{Bmatrix}^T \begin{bmatrix} 1 & 0 & 0 \\ 0 & 1 & 0 \\ 0 & 0 & 1 \end{bmatrix} \begin{Bmatrix} 1 \\ -\sqrt{2} \\ 1 \end{Bmatrix} = 1 \tag{3.173}$$

$$\text{or} \quad \gamma_3^2 m \left(1 + \sqrt{2}\sqrt{2} + 1 \right) = 1, \quad \text{that is,} \quad \gamma_3 = \frac{1}{2\sqrt{m}}. \tag{3.174}$$

Thus the mass normalized third eigenvector is given by

$$\mathbf{x}_3 = \frac{1}{2\sqrt{m}} \begin{Bmatrix} 1 \\ -\sqrt{2} \\ 1 \end{Bmatrix}. \tag{3.175}$$

Combining the three eigenvectors, the mass normalized undamped modal matrix is now given by

$$\mathbf{X} = [\mathbf{x}_1, \mathbf{x}_2, \mathbf{x}_3] = \frac{1}{2\sqrt{m}} \begin{bmatrix} 1 & \sqrt{2} & 1 \\ \sqrt{2} & 0 & -\sqrt{2} \\ 1 & -\sqrt{2} & 1 \end{bmatrix}. \tag{3.176}$$

Here the modal matrix turns out to be symmetric. But in general this is not the case.

IV. *Obtain the modal damping factors:* The damping matrix in the modal coordinate can be obtained from (3.97) as

$$\mathbf{C}' = \mathbf{X}^T \mathbf{C} \mathbf{X} = \frac{1}{2\sqrt{m}} \begin{bmatrix} 1 & \sqrt{2} & 1 \\ \sqrt{2} & 0 & -\sqrt{2} \\ 1 & -\sqrt{2} & 1 \end{bmatrix}^T \begin{bmatrix} c & 0 & 0 \\ 0 & c & 0 \\ 0 & 0 & c \end{bmatrix} \tag{3.177}$$

$$\frac{1}{2\sqrt{m}} \begin{bmatrix} 1 & \sqrt{2} & 1 \\ \sqrt{2} & 0 & -\sqrt{2} \\ 1 & -\sqrt{2} & 1 \end{bmatrix} = \frac{c}{m} \mathbf{I}.$$

Therefore

$$2\zeta_j\omega_j = \frac{c}{m} \quad \text{or} \quad \zeta_j = \frac{c}{2m\omega_j}. \tag{3.178}$$

Since ω_j becomes bigger for higher modes, modal damping gets smaller, i.e., higher modes are less damped.

V. *Response due to applied loading:* The applied loading $\mathbf{f}(t) = \{f(t), 0, 0\}^T$ where $f(t) = 1 - U(t - t_0)$ with $t_0 = \frac{2\pi}{\omega_1}$. In the Laplace domain

$$\bar{f}(s) = \mathcal{L}\left[1 - U(t - t_0)\right] = 1 - \frac{e^{-st_0}}{s}. \tag{3.179}$$

Thus, the term $\mathbf{x}_j^T \bar{\mathbf{f}}(s)$ can be obtained as

$$\mathbf{x}_j^T \bar{\mathbf{f}}(s) = \left\{ \begin{matrix} x_{1j} \\ x_{2j} \\ x_{3j} \end{matrix} \right\}^T \left\{ \begin{matrix} \bar{f}(s) \\ 0 \\ 0 \end{matrix} \right\} = x_{1j}\left(1 - \frac{e^{-st_0}}{s}\right) \quad \forall j. \tag{3.180}$$

Since the initial conditions are zero, the dynamic response in the Laplace domain can be obtained from (3.134) as

$$\bar{\mathbf{q}}(s) = \sum_{j=1}^{3}\left\{\frac{x_{1j}\left(1 - \frac{e^{-st_0}}{s}\right)}{s^2 + 2s\zeta_j\omega_j + \omega_j^2}\right\}\mathbf{x}_j = \sum_{j=1}^{3}\left\{\frac{x_{1j}\left(s - e^{-st_0}\right)}{s\left(s^2 + 2s\zeta_j\omega_j + \omega_j^2\right)}\right\}\mathbf{x}_j. \tag{3.181}$$

In the frequency domain, the response is given by

$$\bar{\mathbf{q}}(i\omega) = \sum_{j=1}^{3}\left\{\frac{x_{1j}\left(i\omega - e^{-i\omega t_0}\right)}{i\omega\left(-\omega^2 + 2i\omega\zeta_j\omega_j + \omega_j^2\right)}\right\}\mathbf{x}_j. \tag{3.182}$$

For the numerical calculations we assume $m = 1$, $k = 1$ and $c = 0.2$. Using these values, from (3.182), the absolute value of the frequency-domain response of the three masses is plotted in Figure 3.6. The three peaks in the diagram correspond to the three natural frequencies of the system. The time-domain response can be obtained by evaluating the convolution integral in (3.142) and substituting $a_j(t)$ in (3.137). In practice, usually numerical integration methods are used to evaluate this integral. For this problem a closed-from expression can be obtained. We have

$$\mathbf{x}_j^T \mathbf{f}(\tau) = x_{1j}f(\tau). \tag{3.183}$$

From Figure 3.5 it can be noted that

$$f(\tau) = \begin{cases} 1 & \text{if} \quad \tau < t_0, \\ 0 & \text{if} \quad \tau > t_0. \end{cases} \tag{3.184}$$

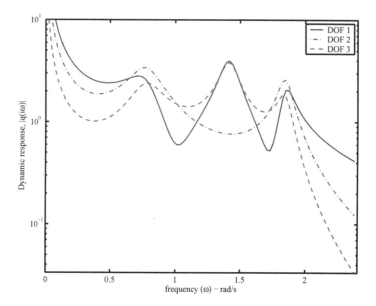

Figure 3.6 Absolute value of the frequency-domain response of the three masses due to applied step loading at first DOF.

Because of this, the limit of the integral in (3.142) can be changed as

$$a_j(t) = \int_0^t \frac{1}{\omega_{d_j}} \mathbf{x}_j^T \mathbf{f}(\tau) e^{-\zeta_j \omega_j (t-\tau)} \sin\left(\omega_{d_j}(t-\tau)\right) d\tau$$

$$= \int_0^{t_0} \frac{1}{\omega_{d_j}} x_{1j} e^{-\zeta_j \omega_j (t-\tau)} \sin\left(\omega_{d_j}(t-\tau)\right) d\tau \qquad (3.185)$$

$$= \frac{x_{1j}}{\omega_{d_j}} \int_0^{t_0} e^{-\zeta_j \omega_j (t-\tau)} \sin\left(\omega_{d_j}(t-\tau)\right) d\tau.$$

By making a substitution $\tau' = t - \tau$, this integral can be evaluated as

$$a_j(t) = \frac{x_{1j}}{\omega_{d_j}} \frac{e^{-\zeta_j \omega_j t}}{\omega_j^2} \left\{ \alpha_j \sin\left(\omega_{d_j} t\right) + \beta_j \cos\left(\omega_{d_j} t\right) \right\} \qquad (3.186)$$

where $\quad \alpha_j = \left\{ \omega_{d_j} \sin\left(\omega_{d_j} t_0\right) + \zeta_j \omega_j \cos\left(\omega_{d_j} t_0\right) \right\} e^{\zeta_j \omega_j t_0} - \zeta_j \omega_j$
$$(3.187)$$

and $\quad \beta_j = \left\{ \omega_{d_j} \cos\left(\omega_{d_j} t_0\right) - \zeta_j \omega_j \sin\left(\omega_{d_j} t_0\right) \right\} e^{\zeta_j \omega_j t_0} - \omega_{d_j}. \quad (3.188)$

The time-domain responses of the three masses obtained using Equation 3.137 are shown in Figure 3.7. From the diagram observe that, because the forcing is applied to the first mass (DOF 1), it moves earlier and comparatively more than the other two masses.

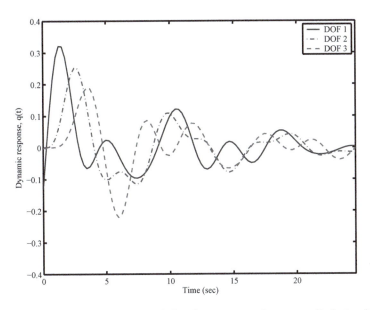

Figure 3.7 Time domain response of the three masses due to applied step loading at first DOF.

VI. *Response due to initial displacement:* When $\mathbf{q}_0 = \{0, 1, 0\}^T$ we have

$$\mathbf{x}_j^T \mathbf{C} \mathbf{q}_0 = \begin{Bmatrix} x_{1j} \\ x_{2j} \\ x_{3j} \end{Bmatrix}^T \begin{bmatrix} c & 0 & 0 \\ 0 & c & 0 \\ 0 & 0 & c \end{bmatrix} \cdot \begin{Bmatrix} 0 \\ 1 \\ 0 \end{Bmatrix} = x_{2j} c \quad \forall j. \tag{3.189}$$

Similarly

$$\mathbf{x}_j^T \mathbf{M} \mathbf{q}_0 = x_{2j} m \quad \forall j. \tag{3.190}$$

The dynamic response in the Laplace domain can be obtained from (3.134) as

$$\begin{aligned}
\bar{\mathbf{q}}(s) &= \sum_{j=1}^{3} \left\{ \frac{x_{2j} c}{s^2 + 2s\zeta_j \omega_j + \omega_j^2} + \frac{x_{2j} m s}{s^2 + 2s\zeta_j \omega_j + \omega_j^2} \right\} \mathbf{x}_j \\
&= \sum_{j=1}^{3} x_{2j} \left\{ \frac{c + m s}{s^2 + 2s\zeta_j \omega_j + \omega_j^2} \right\} \mathbf{x}_j.
\end{aligned} \tag{3.191}$$

From (3.178) note that $c = 2\zeta_j \omega_j m$. Substituting this in the above equation we have

$$\bar{\mathbf{q}}(s) = \sum_{j=1}^{3} x_{2j} m \left\{ \frac{2\zeta_j \omega_j + s}{s^2 + 2s\zeta_j \omega_j + \omega_j^2} \right\} \mathbf{x}_j. \tag{3.192}$$

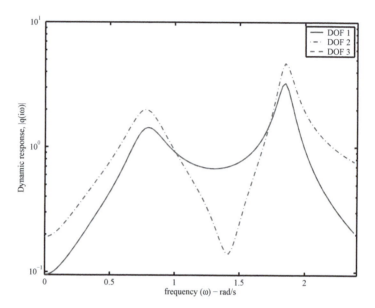

Figure 3.8 Absolute value of the frequency-domain response of the three masses due to unit initial displacement at second DOF.

In the frequency domain the response is given by

$$\bar{\mathbf{q}}(i\omega) = \sum_{j=1}^{3} x_{2j} m \left\{ \frac{2\zeta_j \omega_j + i\omega}{-\omega^2 + 2i\omega\zeta_j\omega_j + \omega_j^2} \right\} \mathbf{x}_j. \tag{3.193}$$

Figure 3.8 shows the responses of the three masses. Note that the second peak is "missing" and the responses of the first and the third masses are exactly the same. This is due to the fact that in the second mode of vibration the middle mass remains stationary while the two other masses move equal amount in the opposite directions (see the second mode in [3.168]). The time-domain response can be obtained by directly taking the inverse Laplace transform of (3.192) as

$$\mathbf{q}(t) = \mathcal{L}^{-1}[\bar{\mathbf{q}}(s)] = \sum_{j=1}^{3} x_{2j} m \mathcal{L}^{-1} \left[\frac{2\zeta_j\omega_j + s}{s^2 + 2s\zeta_j\omega_j + \omega_j^2} \right] \mathbf{x}_j$$

$$= \sum_{j=1}^{3} x_{2j} m e^{-\zeta_j\omega_j t} \cos\left(\omega_{d_j} t\right) \mathbf{x}_j. \tag{3.194}$$

This expression is plotted in 3.9. Observe that initial displacement of the second mass is unity, which verify that the initial condition has been applied correctly. Because of the symmetry of the system the displacements of the two other masses are identical.

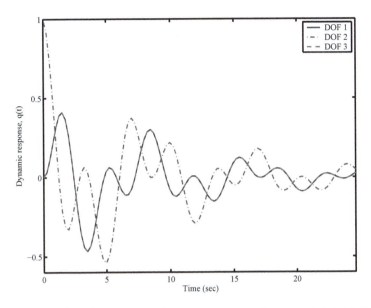

Figure 3.9 Time domain response of the three masses due to unit initial displacement at second DOF.

VII. *Response due to initial velocity:* When $\dot{\mathbf{q}}_0 = \{0, 1, 1\}^T$ we have

$$\mathbf{x}_j^T \mathbf{M} \dot{\mathbf{q}}_0 = \begin{Bmatrix} x_{1j} \\ x_{2j} \\ x_{3j} \end{Bmatrix}^T \begin{bmatrix} m & 0 & 0 \\ 0 & m & 0 \\ 0 & 0 & m \end{bmatrix} \begin{Bmatrix} 0 \\ 1 \\ 1 \end{Bmatrix} = (x_{2j} + x_{3j})\, m \quad \forall j. \quad (3.195)$$

The dynamic response in the Laplace domain can be obtained from Equation (3.134) as

$$\bar{\mathbf{q}}(s) = \sum_{j=1}^{3} \left\{ \frac{(x_{2j} + x_{3j})\, m}{s^2 + 2s\zeta_j\omega_j + \omega_j^2} \right\} \mathbf{x}_j. \quad (3.196)$$

The time-domain response can be obtained using the inverse Laplace transform as

$$\mathbf{q}(t) = \sum_{j=1}^{3} (x_{2j} + x_{3j}) \frac{m}{\omega_{d_j}} e^{-\zeta_j\omega_j t} \sin\left(\omega_{d_j} t\right) \mathbf{x}_j. \quad (3.197)$$

The responses of the three masses in the frequency domain and in the time domain are respectively shown in Figures 3.10 and 3.11. In this case all the modes of the system can be observed. Because the initial conditions of the second and the third masses are the same, their initial displacements are close to each other. However, as the time passes the displacements of these two masses start differing from each other.

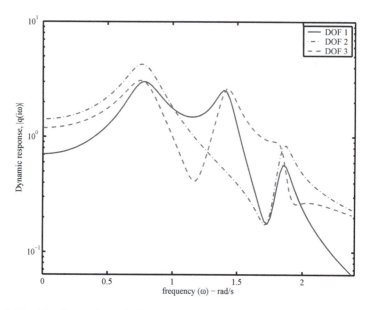

Figure 3.10 Absolute value of the frequency-domain response of the three masses due to unit initial velocity at the second and third DOF.

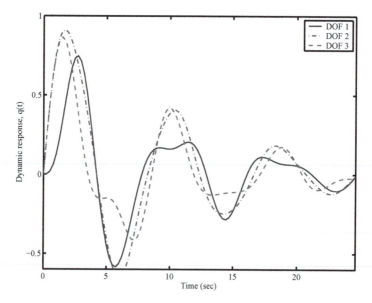

Figure 3.11 Time-domain response of the three masses due to unit initial velocity at the second and third DOF.

3.5 NONPROPORTIONALLY DAMPED SYSTEMS

Modes of proportionally damped systems preserve the simplicity of the real normal modes as in the undamped case. Unfortunately there is no physical reason why a general system should behave like this. In fact practical experience in modal testing shows that most real-life structures do not do so, as they possess complex modes instead of real normal modes. This implies that in general linear systems are nonclassically damped. When the system is nonclassically damped, some or all of the N differential equations in the modal coordinate (3.96) are coupled through the $\mathbf{X}^T \mathbf{C} \mathbf{X}$ term and cannot be reduced to N second-order uncoupled equations. This coupling brings several complication in the system dynamics—the eigenvalues and the eigenvectors no longer remain real and also the eigenvectors do not satisfy the classical orthogonality relationships as given by (3.63) and (3.64).

3.5.1 FREE VIBRATION AND COMPLEX MODES

The complex eigenvalue problem associated with the equation of motion (3.95) can be represented by

$$s_j^2 \mathbf{M} \mathbf{z}_j + s_j \mathbf{C} \mathbf{z}_j + \mathbf{K} \mathbf{z}_j = \mathbf{0} \tag{3.198}$$

where $s_j \in \mathbb{C}$ is the jth eigenvalue and $\mathbf{z}_j \in \mathbb{C}^N$ is the jth eigenvector. The eigenvalues, s_j, are the roots of the characteristic polynomial

$$\det \left[s^2 \mathbf{M} + s \mathbf{C} + \mathbf{K} \right] = 0. \tag{3.199}$$

The order of the polynomial is $2N$ and if the roots are complex they appear in complex conjugate pairs. The methods for solving this kind of complex problem follow mainly two routes, the state-space method and the methods in configuration space or "N-space." Brief discussions on these two approaches are taken up in the following subsections.

3.5.1.1 The State-Space Method

The state-space method is based on transforming the N second-order coupled equations into a set of $2N$ first-order coupled equations by augmenting the displacement response vectors with the velocities of the corresponding coordinates. We can write Equation 3.95 together with a trivial equation $\mathbf{M}\dot{\mathbf{q}}(t) - \mathbf{M}\dot{\mathbf{q}}(t) = 0$ in a matrix form as

$$\begin{bmatrix} \mathbf{C} & \mathbf{M} \\ \mathbf{M} & \mathbf{O} \end{bmatrix} \begin{Bmatrix} \dot{\mathbf{q}}(t) \\ \ddot{\mathbf{q}}(t) \end{Bmatrix} + \begin{bmatrix} \mathbf{K} & \mathbf{O} \\ \mathbf{O} & -\mathbf{M} \end{bmatrix} \begin{Bmatrix} \mathbf{q}(t) \\ \dot{\mathbf{q}}(t) \end{Bmatrix} = \begin{Bmatrix} \mathbf{f}(t) \\ \mathbf{0} \end{Bmatrix} \tag{3.200}$$

or $\quad \mathbf{A}\,\dot{\mathbf{u}}(t) + \mathbf{B}\,\mathbf{u}(t) = \mathbf{r}(t) \tag{3.201}$

where

$$\mathbf{A} = \begin{bmatrix} \mathbf{C} & \mathbf{M} \\ \mathbf{M} & \mathbf{O} \end{bmatrix} \in \mathbb{R}^{2N \times 2N}, \quad \mathbf{B} = \begin{bmatrix} \mathbf{K} & \mathbf{O} \\ \mathbf{O} & -\mathbf{M} \end{bmatrix} \in \mathbb{R}^{2N \times 2N},$$

$$\mathbf{u}(t) = \left\{ \begin{matrix} \mathbf{q}(t) \\ \dot{\mathbf{q}}(t) \end{matrix} \right\} \in \mathbb{R}^{2N}, \quad \text{and} \quad \mathbf{r}(t) = \left\{ \begin{matrix} \mathbf{f}(t) \\ \mathbf{0} \end{matrix} \right\} \in \mathbb{R}^{2N}. \tag{3.202}$$

In the above equation $\mathbf{O}$ is the $N \times N$ null matrix. This form of equation of motion is also known as the "Duncan form." The eigenvalue problem associated with (3.201) can be expressed as

$$s_j \mathbf{A} \boldsymbol{\phi}_j + \mathbf{B} \boldsymbol{\phi}_j = \mathbf{0} \tag{3.203}$$

where $s_j \in \mathbb{C}$ is the j-th eigenvalue and $\boldsymbol{\phi}_j \in \mathbb{C}^{2N}$ is the j-th eigenvector. This eigenvalue problem is similar to the undamped eigenvalue problem (3.55) except (a) the dimension of the matrices are $2N$ as opposed to N and (b) the matrices are not positive definite. Because of (a) the computational cost to obtain the eigensolutions of (3.203) is much higher compared to the undamped eigensolutions and due to (b) the eigensolutions in general become complex valued. From a phenomenological point of view, this implies that the modes are not synchronous, i.e., there is a "phase lag" so that different degrees of freedom do not simultaneously reach to their corresponding 'peaks' and "troughs." Thus, both from computational and conceptual point of view, complex modes significantly complicates the problem and in practice they are often avoided. From the expression of $\mathbf{u}(t)$ in (3.202), the state-space complex eigenvectors $\boldsymbol{\phi}_j$ can be related to the jth eigenvector of the second-order system as

$$\boldsymbol{\phi}_j = \left\{ \begin{matrix} \mathbf{z}_j \\ s_j \mathbf{z}_j \end{matrix} \right\}. \tag{3.204}$$

Since $\mathbf{A}$ and $\mathbf{B}$ are real matrices, taking complex conjugate $((\bullet)^*$ denotes complex conjugation) of the eigenvalue (3.203) it is trivial to see that

$$s_j^* \mathbf{A} \boldsymbol{\phi}_j^* + \mathbf{B} \boldsymbol{\phi}_j^* = \mathbf{0}. \tag{3.205}$$

This implies that the eigensolutions must appear in complex conjugate pairs. For convenience arrange the eigenvalues and the eigenvectors so that

$$s_{j+N} = s_j^* \tag{3.206}$$

$$\boldsymbol{\phi}_{j+N} = \boldsymbol{\phi}_j^*, \quad j = 1, 2, \cdots, N \tag{3.207}$$

Like the real normal modes, complex modes in the state-space also satisfy orthogonal relationships over the $\mathbf{A}$ and $\mathbf{B}$ matrices. For distinct eigenvalues it is easy to show that

$$\boldsymbol{\phi}_j^T \mathbf{A} \boldsymbol{\phi}_k = 0 \quad \text{and} \quad \boldsymbol{\phi}_j^T \mathbf{B} \boldsymbol{\phi}_k = 0; \quad \forall j \neq k. \tag{3.208}$$

Premultiplying Equation (3.203) by $\mathbf{y}_j^T$ one obtains

$$\phi_j^T \mathbf{B} \phi_j = -s_j \phi_j^T \mathbf{A} \phi_j. \tag{3.209}$$

The eigenvectors may be normalized so that

$$\phi_j^T \mathbf{A} \phi_j = \gamma_j \tag{3.210}$$

where $\gamma_j \in \mathbb{C}$ is the normalization constant. In view of the expressions of ϕ_j in (3.204) the above relationship can be expressed in terms of the eigensolutions of the second-order system as

$$\mathbf{z}_j^T [2s_j \mathbf{M} + \mathbf{C}] \mathbf{z}_j = \gamma_j. \tag{3.211}$$

There are several ways in which the normalization constants can be selected. The one that is most consistent with traditional modal analysis practice, is to choose $\gamma_j = 2s_j$. Observe that this degenerates to the familiar mass normalization relationship $\mathbf{z}_j^T \mathbf{M} \mathbf{z}_j = 1$ when the damping is zero.

3.5.1.2   Approximate Methods in the Configuration Space

It has been pointed out that the state-space approach toward the solution of equation of motion in the context of linear structural dynamics is not only computationally expensive but also fails to provide the physical insight which modal analysis in the configuration space or N-space offers.
Light damping assumption:

Assuming that the damping is light, a simple *first-order perturbation method* is used to obtain complex modes and frequencies in terms of undamped modes and frequencies.

The undamped modes form a *complete* set of vectors so that each complex mode $\mathbf{z}_j$ can be expressed as a linear combination of $\mathbf{x}_j$. Assume that

$$\mathbf{z}_j = \sum_{k=1}^{N} \alpha_k^{(j)} \mathbf{x}_k \tag{3.212}$$

where $\alpha_k^{(j)}$ are complex constants which we want to determine. Since damping is assumed light, $\alpha_k^{(j)} \ll 1, \forall j \neq k$ and $\alpha_j^{(j)} = 1, \forall j$. Suppose the complex natural frequencies are denoted by λ_j, which are related to complex eigenvalues s_j through

$$s_j = \mathrm{i}\lambda_j. \tag{3.213}$$

Substituting s_j and $\mathbf{z}_j$ in the eigenvalue (3.198) we have

$$[-\lambda_j^2 \mathbf{M} + \mathrm{i}\lambda_j \mathbf{C} + \mathbf{K}] \sum_{k=1}^{N} \alpha_k^{(j)} \mathbf{x}_k = \mathbf{0}. \tag{3.214}$$

Premultiplying by $\mathbf{x}_j^T$ and using the orthogonality conditions (3.63) and (3.64), we have

$$-\lambda_j^2 + i\lambda_j \sum_{k=1}^{N} \alpha_k^{(j)} C'_{jk} + \omega_j^2 = 0 \qquad (3.215)$$

where $C'_{jk} = \mathbf{x}_j^T \mathbf{C} \mathbf{x}_k$, is the jkth element of the modal damping matrix $\mathbf{C}'$. Due to small damping assumption we can neglect the product $\alpha_k^{(j)} C'_{jk}, \forall j \neq k$ since they are small compared to $\alpha_j^{(j)} C'_{jj}$. Thus, Equation 3.215 can be approximated as

$$-\lambda_j^2 + i\lambda_j \alpha_j^{(j)} C'_{jj} + \omega_j^2 \approx 0. \qquad (3.216)$$

By solving the quadratic equation we have

$$\lambda_j \approx \pm\omega_j + iC'_{jj}/2. \qquad (3.217)$$

This is the expression of approximate complex natural frequencies. Premultiplying (3.214) by $\mathbf{x}_k^T$ using the orthogonality conditions (3.63) and (3.64) and light damping assumption, it can be shown that

$$\alpha_k^{(j)} \approx \frac{i\omega_j C'_{kj}}{\omega_j^2 - \omega_k^2}, \quad k \neq j. \qquad (3.218)$$

Substituting this in (3.212), approximate complex modes can be given by

$$\mathbf{z}_j \approx \mathbf{x}_j + \sum_{k \neq j}^{N} \frac{i\omega_j C'_{kj} \mathbf{x}_k}{\omega_j^2 - \omega_k^2}. \qquad (3.219)$$

This expression shows that (a) the imaginary parts of complex modes are approximately orthogonal to the real parts, and (b) the "complexity" of the modes will be more if ω_j and ω_k are close, i.e., modes will be significantly complex when the natural frequencies of a system are closely spaced. It should be recalled that the first-order perturbation expressions are only valid when damping is small.

Light non-proportional damping:

Here we assume that the system has light non-proportional damping. Therefore, a higher amount of damping can be used provided it not significantly non-proportional. Equation (3.214) can be rewritten as

$$-\lambda_j^2 \alpha_k^{(j)} + i\lambda_j \left(\alpha_j^{(j)} C'_{ki} + \alpha_k^{(j)} C'_{kk} + \sum_{l \neq k \neq i}^{N} \alpha_j^{(j)} C'_{kl} \right) + \omega_k^2 \alpha_k^{(j)} = 0 \quad (3.220)$$

which leads to

$$\alpha_k^{(j)} \approx -\frac{i\lambda_j C'_{kj}}{\omega_k^2 - \lambda_j^2 + i\lambda_j C'_{kk}}. \qquad (3.221)$$

By replacing ω_k from (3.217) for the k-th set, the denominator appearing in the above expression can be factorized as

$$\omega_k^2 - \lambda_j^2 + i\lambda_j C_{kk}' \approx -(\lambda_j - \lambda_k)(\lambda_j - \lambda_k^*) \tag{3.222}$$

where $(\bullet)^*$ denotes complex conjugation. With the help of this factorization, Equation (3.221) can be expressed as

$$\alpha_k^{(j)} \approx \frac{i\lambda_j C_{kj}'}{(\lambda_j - \lambda_k)(\lambda_j - \lambda_k^*)} \tag{3.223}$$

and consequently from the series sum of (3.212) the approximate expression for the complex modes reads

$$\mathbf{z}_j \approx \mathbf{x}_j + \sum_{k=1}^{N} \frac{i\lambda_j C_{kj}' \mathbf{x}_k}{(\lambda_j - \lambda_k)(\lambda_j - \lambda_k^*)}. \tag{3.224}$$

The approach taken here is very similar to the classical perturbation analysis described before, but the $\alpha_k^{(j)}$ expressed by (3.223) appear to be slightly different. The classical expression for this is given by (3.218), which is equivalent to replacing the complex natural frequencies by undamped natural frequencies in (3.223). Numerical calculations using the $\alpha_k^{(j)}$ described by (3.223) yield more accurate results than those of the classical analysis.

Example 3.4. To illustrate the accuracy of different approximation, consider a two-degree-of-freedom system shown in Figure 3.12. The system matrices for the example taken can be expressed by

$$\mathbf{M} = \begin{bmatrix} 1 & 0 \\ 0 & 1 \end{bmatrix}; \quad \mathbf{C} = \begin{bmatrix} 4 & -4 \\ -4 & 4 \end{bmatrix} \quad \text{and} \quad \mathbf{K} = \begin{bmatrix} 1100 & -100 \\ -100 & 1200 \end{bmatrix} \tag{3.225}$$

Using Equation 3.224, the approximate complex mode shapes can be calculated as

$$\mathbf{z}_1 = \begin{bmatrix} 0.7870 - 0.0661i \\ 0.6287 + 0.1070i \end{bmatrix}; \quad \mathbf{z}_2 = \begin{bmatrix} -0.6406 - 0.1179i \\ 0.7797 - 0.0729i \end{bmatrix}. \tag{3.226}$$

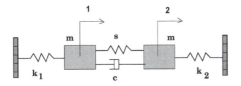

Figure 3.12 A two-degree-of-system, $m = 1$ kg, $k_1 = 1000$ N/m, $s = 100$ N/m $k_2 = 1.1k_1$ and $c = 4.0$ Ns/m.

The corresponding exact mode shapes from the state-space formulation can be obtained as

$$\mathbf{u}_1^e = \begin{bmatrix} 0.7870 - 0.0661\mathrm{i} \\ 0.6443 + 0.1018\mathrm{i} \end{bmatrix}; \quad \mathbf{u}_2^e = \begin{bmatrix} -0.6406 - 0.1179\mathrm{i} \\ 0.7631 - 0.0603\mathrm{i} \end{bmatrix} \quad (3.227)$$

while the classical perturbation method gives

$$\mathbf{u}_1^c = \begin{bmatrix} 0.7870 - 0.0661\mathrm{i} \\ 0.4658 + 0.2362\mathrm{i} \end{bmatrix}; \quad \mathbf{u}_2^c = \begin{bmatrix} -0.6406 - 0.1179\mathrm{i} \\ 0.8866 - 0.3989\mathrm{i} \end{bmatrix}. \quad (3.228)$$

For comparison, the modes are normalized to have the same numerical value in the first element, so that only the second element differs. It is clear that the results obtained from the expression suggested here are closer to the exact values than the classical analysis. Beside this, the expression in (3.224) also indicates a conceptual difference from the classical analysis: unlike there, the "correction terms" no longer remain purely imaginary. This in turn means that the real part of the complex mode shapes are not the undamped mode shapes. This fact is indeed true and can be verified from the exact analysis.

3.5.2 DYNAMIC RESPONSE

Once the complex mode shapes and natural frequencies are obtained (either from the state-space method or from an approximate method), the dynamic response can be obtained using the orthogonality properties of the complex eigenvectors in the state-space. We will derive the expressions for the general dynamic response in the frequency and time domain.

3.5.2.1 Frequency-Domain Analysis

Taking the Laplace transform of the state-space equation of motion (3.201) we have

$$s\mathbf{A}\bar{\mathbf{u}}(s) - \mathbf{A}\mathbf{u}_0 + \mathbf{B}\bar{\mathbf{u}}(s) = \bar{\mathbf{r}}(s) \quad (3.229)$$

where $\bar{\mathbf{u}}(s)$ is the Laplace transform of $\mathbf{u}(t)$, $\mathbf{u}_0$ is the vector of initial conditions in the state-space and $\bar{\mathbf{r}}(s)$ is the Laplace transform of $\mathbf{r}(t)$. From the expressions of $\mathbf{u}(t)$ and $\mathbf{r}(t)$ in (3.202) it is obvious that

$$\bar{\mathbf{u}}(s) = \left\{ \begin{matrix} \bar{\mathbf{q}}(s) \\ s\bar{\mathbf{q}}(s) \end{matrix} \right\} \in \mathbb{C}^{2N}, \quad \bar{\mathbf{r}}(s) = \left\{ \begin{matrix} \bar{\mathbf{f}}(s) \\ \mathbf{0} \end{matrix} \right\} \in \mathbb{C}^{2N} \quad \text{and} \quad \mathbf{u}_0 = \left\{ \begin{matrix} \mathbf{q}_0 \\ \dot{\mathbf{q}}_0 \end{matrix} \right\} \in \mathbb{R}^{2N} \quad (3.230)$$

For distinct eigenvalues the mode shapes $\boldsymbol{\phi}_k, k = 1, 2, \cdots, 2N$ form a complete set of vectors. Therefore, the solution of (3.229) can be expressed in terms of a linear combination of $\boldsymbol{\phi}_k$ as

$$\bar{\mathbf{u}}(s) = \sum_{k=1}^{2N} \beta_k(s)\boldsymbol{\phi}_k. \quad (3.231)$$

We only need to determine the constants $\beta_k(s)$ to obtain the complete solution. Substituting $\bar{\mathbf{u}}(s)$ from (3.231) into (3.229) we have

$$[s\mathbf{A} + \mathbf{B}] \sum_{k=1}^{2N} \beta_k(s)\boldsymbol{\phi}_k = \bar{\mathbf{r}}(s) + \mathbf{A}\mathbf{u}_0. \tag{3.232}$$

Premultiplying by $\boldsymbol{\phi}_j^T$ and using the orthogonality and normalization relationships (3.208)–(3.210), we have

$$\gamma_j (s - s_j) \beta_j(s) = \boldsymbol{\phi}_j^T \{\bar{\mathbf{r}}(s) + \mathbf{A}\mathbf{u}_0\}$$

$$\text{or} \quad \beta_j(s) = \frac{1}{\gamma_j} \frac{\boldsymbol{\phi}_j^T \bar{\mathbf{r}}(s) + \boldsymbol{\phi}_j^T \mathbf{A}\mathbf{u}_0}{s - s_j}. \tag{3.233}$$

Using the expressions of $\mathbf{A}$, $\boldsymbol{\phi}_j$ and $\bar{\mathbf{r}}(s)$ from (3.202), (3.204) and (3.230) the term $\boldsymbol{\phi}_j^T \bar{\mathbf{r}}(s) + \boldsymbol{\phi}_j^T \mathbf{A}\mathbf{u}_0$ can be simplified and $\beta_k(s)$ can be related with the mode shapes of the second-order system as

$$\beta_j(s) = \frac{1}{\gamma_j} \frac{\mathbf{z}_j^T \{\bar{\mathbf{f}}(s) + \mathbf{M}\dot{\mathbf{q}}_0 + \mathbf{C}\mathbf{q}_0 + s\mathbf{M}\mathbf{q}_0\}}{s - s_j}. \tag{3.234}$$

Since we are only interested in the displacement response, we only need to determine the first N rows of (3.231). Using the partition of $\mathbf{u}(s)$ and $\boldsymbol{\phi}_j$ we have

$$\bar{\mathbf{q}}(s) = \sum_{j=1}^{2N} \beta_j(s)\mathbf{z}_j. \tag{3.235}$$

Substituting $\beta_j(s)$ from (3.234) into the preceding equation one has

$$\bar{\mathbf{q}}(s) = \sum_{j=1}^{2N} \frac{1}{\gamma_j} \mathbf{z}_j \frac{\mathbf{z}_j^T \{\bar{\mathbf{f}}(s) + \mathbf{M}\dot{\mathbf{q}}_0 + \mathbf{C}\mathbf{q}_0 + s\mathbf{M}\mathbf{q}_0\}}{s - s_j} \tag{3.236}$$

$$\text{or} \quad \bar{\mathbf{q}}(s) = \mathbf{H}(s) \{\bar{\mathbf{f}}(s) + \mathbf{M}\dot{\mathbf{q}}_0 + \mathbf{C}\mathbf{q}_0 + s\mathbf{M}\mathbf{q}_0\} \tag{3.237}$$

where

$$\mathbf{H}(s) = \sum_{j=1}^{2N} \frac{1}{\gamma_j} \frac{\mathbf{z}_j \mathbf{z}_j^T}{s - s_j} \tag{3.238}$$

is the transfer function matrix. Recalling that the eigensolutions appear in complex conjugate pairs, using (3.206), the transfer function matrix (3.238) can be expanded as

$$\mathbf{H}(s) = \sum_{j=1}^{2N} \frac{\mathbf{z}_j \mathbf{z}_j^T}{\gamma_j(s - s_j)} = \sum_{j=1}^{N} \left[\frac{\mathbf{z}_j \mathbf{z}_j^T}{\gamma_j(s - s_j)} + \frac{\mathbf{z}_j^* \mathbf{z}_j^{*T}}{\gamma_j^*(s - s_j^*)} \right]. \tag{3.239}$$

The transfer function matrix is often expressed in terms of the complex natural frequencies λ_j. Substituting $s_j = i\lambda_j$ in the preceding expression we have

$$\mathbf{H}(s) = \sum_{j=1}^{N} \left[\frac{\mathbf{z}_j \mathbf{z}_j^T}{\gamma_j(s - i\lambda_j)} + \frac{\mathbf{z}_j^* \mathbf{z}_j^{*^T}}{\gamma_j^*(s + i\lambda_j^*)} \right] \quad \text{and} \quad \gamma_j = \mathbf{z}_j^T \left[2i\lambda_j \mathbf{M} + \mathbf{C} \right] \mathbf{z}_j.$$

(3.240)

It can be shown that the transfer function matrix $\mathbf{H}(s)$ in (3.240) reduces to its equivalent expression for the undamped case in (3.85). In the undamped limit $\mathbf{C} = 0$. This results $\lambda_j = \omega_j = \lambda_j^*$ and $\mathbf{z}_j = \mathbf{x}_j = \mathbf{z}_j^*$. In view of the mass normalization relationship we also have $\gamma_j = 2i\omega_j$. Consider a typical term in (3.240):

$$\left[\frac{\mathbf{z}_j \mathbf{z}_j^T}{\gamma_j(s - i\lambda_j)} + \frac{\mathbf{z}_j^* \mathbf{z}_j^{*^T}}{\gamma_j^*(s + i\lambda_j^*)} \right] = \left[\frac{1}{2i\omega_j} \frac{1}{i\omega - i\omega_j} + \frac{1}{-2i\omega_j} \frac{1}{i\omega + i\omega_j} \right] \mathbf{x}_j \mathbf{x}_j^T$$

$$= \frac{1}{2i^2\omega_j} \left[\frac{1}{\omega - \omega_j} - \frac{1}{\omega + \omega_j} \right] \mathbf{x}_j \mathbf{x}_j^T$$

$$= -\frac{1}{2\omega_j} \left[\frac{\omega + \omega_j - \omega + \omega_j}{(\omega - \omega_j)(\omega + \omega_j)} \right] \mathbf{x}_j \mathbf{x}_j^T = -\frac{1}{2\omega_j} \left[\frac{2\omega_j}{\omega^2 - \omega_j^2} \right] \mathbf{x}_j \mathbf{x}_j^T = \frac{\mathbf{x}_j \mathbf{x}_j^T}{\omega_j^2 - \omega^2}.$$

(3.241)

Note that this term was derived before for the transfer function matrix of the undamped system in (3.85). Therefore, Equation 3.237 is the most general expression of the dynamic response of damped linear dynamic systems. In a similar manner, it can also be verified that when the system is proportionally damped, (3.237) reduces to (3.133) as expected.

3.5.2.2   Time-Domain Analysis

Combining Equations (3.237) and (3.240) we have

$$\bar{\mathbf{q}}(s) = \sum_{j=1}^{N} \left\{ \frac{\mathbf{z}_j^T \bar{\mathbf{f}}(s) + \mathbf{z}_j^T \mathbf{M} \dot{\mathbf{q}}_0 + \mathbf{z}_j^T \mathbf{C} \mathbf{q}_0 + s \mathbf{z}_j^T \mathbf{M} \mathbf{q}_0}{\gamma_j(s - i\lambda_j)} \mathbf{z}_j + \right.$$

$$\left. \frac{\mathbf{z}_j^{*^T} \bar{\mathbf{f}}(s) + \mathbf{z}_j^{*^T} \mathbf{M} \dot{\mathbf{q}}_0 + \mathbf{z}_j^{*^T} \mathbf{C} \mathbf{q}_0 + s \mathbf{z}_j^{*^T} \mathbf{M} \mathbf{q}_0}{\gamma_j^*(s + i\lambda_j^*)} \mathbf{z}_j^* \right\}. \quad (3.242)$$

From the table of Laplace transforms we know that

$$\mathcal{L}^{-1} \left[\frac{1}{s - a} \right] = e^{at} \quad \text{and} \quad \mathcal{L}^{-1} \left[\frac{s}{s - a} \right] = a e^{at}, \quad t > 0. \quad (3.243)$$

Taking the inverse Laplace transform of (3.242), dynamic response in the time domain can be obtained as

$$\mathbf{q}(t) = \mathcal{L}^{-1}\left[\bar{\mathbf{q}}(s)\right] = \sum_{j=1}^{N} \frac{1}{\gamma_j} a_{1_j}(t)\mathbf{z}_j + \frac{1}{\gamma_j^*} a_{2_j}(t)\mathbf{z}_j^* \qquad (3.244)$$

where

$$a_{1_j}(t) = \mathcal{L}^{-1}\left[\frac{\mathbf{z}_j^T \bar{\mathbf{f}}(s)}{s - i\lambda_j}\right]$$

$$+ \mathcal{L}^{-1}\left[\frac{1}{s - i\lambda_j}\right](\mathbf{z}_j^T \mathbf{M}\dot{\mathbf{q}}_0 + \mathbf{z}_j^T \mathbf{C}\mathbf{q}_0) + \mathcal{L}^{-1}\left[\frac{s}{s - i\lambda_j}\right]\mathbf{z}_j^T \mathbf{M}\mathbf{q}_0 \quad (3.245)$$

$$= \int_0^t e^{i\lambda_j(t-\tau)}\mathbf{z}_j^T \mathbf{f}(\tau)d\tau + e^{i\lambda_j t}\left(\mathbf{z}_j^T \mathbf{M}\dot{\mathbf{q}}_0 + \mathbf{z}_j^T \mathbf{C}\mathbf{q}_0 i\lambda_j \mathbf{z}_j^T \mathbf{M}\mathbf{q}_0\right),$$

$$\text{for} \quad t > 0$$

and similarly

$$a_{2_j}(t) = \mathcal{L}^{-1}\left[\frac{\mathbf{z}_j^{*T} \bar{\mathbf{f}}(s)}{s + i\lambda_j}\right] + \mathcal{L}^{-1}\left[\frac{1}{s + i\lambda_j}\right]\left(\mathbf{z}_j^{*T} \mathbf{M}\dot{\mathbf{q}}_0 + \mathbf{z}_j^{*T} \mathbf{C}\mathbf{q}_0\right)$$

$$+ \mathcal{L}^{-1}\left[\frac{s}{s + i\lambda_j}\right]\mathbf{z}_j*^T \mathbf{M}\mathbf{q}_0$$

$$= \int_0^t e^{-i\lambda_j(t-\tau)}\mathbf{z}_j^{*T} \mathbf{f}(\tau)d\tau + e^{-i\lambda_j t}\left(\mathbf{z}_j^{*T} \mathbf{M}\dot{\mathbf{q}}_0 + \mathbf{z}_j^{*T} \mathbf{C}\mathbf{q}_0 - i\lambda_j \mathbf{z}_j^{*T} \mathbf{M}\mathbf{q}_0\right),$$

$$\text{for} \quad t > 0.$$

$$(3.246)$$

Example 3.5. Figure 3.13 shows a 3-DOF spring-mass system. This system is identical to the one used in example 3.3, except that the damper attached with the middle block is now disconnected.

1. Show that in general the system possesses complex modes.
2. Obtain approximate expressions for complex natural frequencies (using the first-order perturbation method).

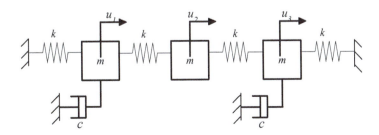

Figure 3.13   3-DOF damped spring-mass system.

Solution: The mass and the stiffness matrices of the system are same as in example 3.3 (given in [3.146]). The damping matrix is clearly given by

$$\mathbf{C} = \begin{bmatrix} c & 0 & 0 \\ 0 & 0 & 0 \\ 0 & 0 & c \end{bmatrix}. \tag{3.247}$$

From this (and also from observation) it clear that the damping matrix is neither proportional to the mass matrix nor it is proportional to the stiffness matrix. Therefore, it is *likely* that the system will not have classical normal mode. It order to be sure we need to check if Caughey and O'Kelly's [26] criteria, i.e., $\mathbf{CM}^{-1}\mathbf{K} = \mathbf{KM}^{-1}\mathbf{C}$, is satisfied or not. Since $\mathbf{M} = m\mathbf{I}$ a diagonal matrix, $\mathbf{M}^{-1} = \frac{1}{m}\mathbf{I}$. Recall that for any matrix $\mathbf{A}$, $\mathbf{IA} = \mathbf{AI} = \mathbf{A}$. Using the system matrices we have

$$\mathbf{CM}^{-1}\mathbf{K} = \mathbf{C}\frac{1}{m}\mathbf{IK} = \frac{1}{m}\mathbf{CK} = \frac{ck}{m}\begin{bmatrix} 1 & 0 & 0 \\ 0 & 0 & 0 \\ 0 & 0 & 1 \end{bmatrix}\begin{bmatrix} 2 & -1 & 0 \\ -1 & 2 & -1 \\ 0 & -1 & 2 \end{bmatrix}$$

$$= \frac{ck}{m}\begin{bmatrix} 2 & -1 & 0 \\ 0 & 0 & 0 \\ 0 & -1 & 2 \end{bmatrix} \tag{3.248}$$

and $\quad \mathbf{KM}^{-1}\mathbf{C} = \mathbf{K}\frac{1}{m}\mathbf{IC} = \frac{1}{m}\mathbf{KC} = \frac{ck}{m}\begin{bmatrix} 2 & -1 & 0 \\ -1 & 2 & -1 \\ 0 & -1 & 2 \end{bmatrix}\begin{bmatrix} 1 & 0 & 0 \\ 0 & 0 & 0 \\ 0 & 0 & 1 \end{bmatrix}$

$$\tag{3.249}$$

$$= \frac{ck}{m}\begin{bmatrix} 2 & 0 & 0 \\ -1 & 0 & -1 \\ 0 & 0 & 2 \end{bmatrix}.$$

It is clear that $\mathbf{CM}^{-1}\mathbf{K} \neq \mathbf{KM}^{-1}\mathbf{C}$, that is, Caughey and O'Kelly's [26] condition is not satisfied by the system matrices. This confirms that the system do not possess classical normal modes but has complex modes.

Exact complex modes of the system can be obtained using the state-space approach outlined in Section 3.5.1.1. For a 3-DOF system we need to solve an eigenvalue problem of the order six. Here we will obtain approximate natural frequencies using the first-order perturbation method described in 3.5.1.2. Using the undamped modal matrix in (3.176), the damping matrix in the modal coordinated can be obtained as

$$\mathbf{C}' = \mathbf{X}^T \mathbf{C} \mathbf{X} = \frac{1}{2\sqrt{m}} \begin{bmatrix} 1 & \sqrt{2} & 1 \\ \sqrt{2} & 0 & -\sqrt{2} \\ 1 & -\sqrt{2} & 1 \end{bmatrix}^T \begin{bmatrix} c & 0 & 0 \\ 0 & 0 & 0 \\ 0 & 0 & c \end{bmatrix} \frac{1}{2\sqrt{m}}$$

$$\begin{bmatrix} 1 & \sqrt{2} & 1 \\ \sqrt{2} & 0 & -\sqrt{2} \\ 1 & -\sqrt{2} & 1 \end{bmatrix}$$

$$= \frac{c}{4m} \begin{bmatrix} 1 & \sqrt{2} & 1 \\ \sqrt{2} & 0 & -\sqrt{2} \\ 1 & -\sqrt{2} & 1 \end{bmatrix} \begin{bmatrix} 1 & 0 & 0 \\ 0 & 0 & 0 \\ 0 & 0 & 1 \end{bmatrix} \begin{bmatrix} 1 & \sqrt{2} & 1 \\ \sqrt{2} & 0 & -\sqrt{2} \\ 1 & -\sqrt{2} & 1 \end{bmatrix} \quad (3.250)$$

$$= \frac{c}{4m} \begin{bmatrix} 1 & \sqrt{2} & 1 \\ \sqrt{2} & 0 & -\sqrt{2} \\ 1 & -\sqrt{2} & 1 \end{bmatrix} \begin{bmatrix} 1 & \sqrt{2} & 1 \\ 0 & 0 & 0 \\ 1 & -\sqrt{2} & 1 \end{bmatrix} = \frac{c}{4m} \begin{bmatrix} 2 & 0 & 2 \\ 0 & 4 & 0 \\ 2 & 0 & 2 \end{bmatrix}$$

$$= \frac{c}{m} \begin{bmatrix} 1/2 & 0 & 1/2 \\ 0 & 1 & 0 \\ 1/2 & 0 & 1/2 \end{bmatrix}.$$

Notice that unlike example 3.3, $\mathbf{C}'$ is not a diagonal matrix, that is, the equation of motion in the modal coordinates are coupled through the off-diagonal terms of the $\mathbf{C}'$ matrix. Approximate complex natural frequencies can be obtained from (3.217) as

$$\lambda_1 \approx \pm\omega_1 + iC'_{11}/2 = \pm\sqrt{\left(2 - \sqrt{2}\right)\alpha} + i\frac{c}{4m}, \quad (3.251)$$

$$\lambda_2 \approx \pm\omega_2 + iC'_{22}/2 = \pm\sqrt{2\alpha} + i\frac{c}{2m} \quad (3.252)$$

$$\text{and} \quad \lambda_3 \approx \pm\omega_3 + iC'_{33}/2 = \pm\sqrt{\left(2 + \sqrt{2}\right)\alpha} + i\frac{c}{4m}. \quad (3.253)$$

In the above equations, undamped eigenvalues obtained in example 3.3 ([3.152]) have been used. The second complex mode is most heavily damped (as the imaginary part is twice compared to the other two modes). This is because in the second mode the middle mass is stationary (look at the second mode shape in [3.165]) while the two "damped" masses move maximum distance away from it. This causes maximum "stretch" to both the dampers and results in maximum damping in this mode.

3.6 SUMMARY

Dynamics of single and multiple-degree-of-freedom undamped and viscously damped systems were discussed. The basic concepts of the natural frequency and the damping factor of a damped SDOF system were introduced. Frequency response and impulse response functions of single-degree-of-freedom undamped and viscously damped systems were derived. The nature of the dynamic response of viscously damped single-degree-of-freedom system is governed by the value of the damping factor. The expressions of general dynamic response due to arbitrary forcing function and initial conditions were derived for single-degree-of-freedom undamped and damped systems. These expressions were then generalized to multiple-degree-of-freedom undamped and proportionally damped systems using the eigenvalues, eigenvectors and modal damping factors. It was shown that the classical modal analysis can be used to obtain the dynamic response of undamped and proportionally damped systems in a similar manner.

The concept of proportional damping was investigated in details. Proportional damping is the most common approach to model dissipative forces in complex engineering structures and it has been used in various dynamic problems for more than ten decades. One of the main limitations of the mass and stiffness proportional damping approximation comes from the fact that the arbitrary variation of damping factors with respect to the vibration frequency cannot be modeled accurately by using this approach. Experimental results, however, suggest that damping factors can vary with frequency. In this chapter a new generalized proportional damping model is discussed in order to capture the frequency-variation of the damping factors accurately. The generalized proportional damping expresses the damping matrix in terms of smooth continuous functions involving specially arranged mass and stiffness matrices so that the system still possesses classical normal modes. This enables one to model variations in the modal damping factors with respect to the frequency in a simplified manner.

Dynamic analysis of general viscously damped systems requires the calculation of complex natural frequencies and complex modes. This can be achieved in a straightforward, but computational expensive, manner by using the state-space approach. A symmetric state-space formulation is outlined in this chapter. The complex eigenvectors in the state-space satisfy orthogonality relationships, similar to that of the classical normal modes in the configuration space. By exploiting the orthogonality relationships, it is possible to extend the modal analysis to obtain the dynamic response using the modal superposition like undamped or proportionally damped systems. The modes however are complex modes and require the solution of a general quadratic eigenvalue problem. A simple perturbation based method based on the light damping assumption is given.

The studies taken up in this chapter gives the necessary platform to discuss more advanced topics for digital twin of dynamic systems. In the next chapter we begin the discussion on stochastic analysis of dynamic systems.

4 Stochastic Analysis

4.1 PROBABILITY THEORY

4.1.1 PROBABILITY SPACE

A trial is a term used to describe the observation of phenomena. Sample space constitute all possible outcome of a trial and can be denoted as Θ. An event θ is defined as a subset of sample space of the given trial Θ. The non-occurrence of the events results in empty set ϕ. If P denotes the probability measure and $\mathcal{F}$ denotes the non-empty collection of subsets holding the following conditions: (1) The empty set phi is in $\mathcal{F}$, (2) If A is in $\mathcal{F}$, then so is the complement of A, and (3)If $A_i, i = 1, 2, 3, ..$ is a sequence of elements of $\mathcal{F}$, then the union of A_i in $\mathcal{F}$, then the construction of probability space is denoted by the notion of triple $(\Theta, \mathcal{F}, P)$

4.1.2 RANDOM VARIABLE

Random variable X can be defined as a mapping such that $X: \Theta \to \mathbb{R}$ on the space $(\Theta, \mathcal{F}, P)$. Consider a realization of random variable X, $x = (x_1, x_2, ..., x_N) : \Omega_x \to \mathbb{R}$, with probability measure of $P(x)$ and cumulative distribution function $F_x(X) = P(x \leq X)$, where P denotes the probability and Ω_x denotes the probability space of x, then probability density function is denoted as $p_x, p_x = dF_x/dx$. Further, for a given continuous function $g(X) = X^l$, lth statistical moment can be defined as $\mathbb{E}[X^l] \equiv \int_{-\infty}^{\infty} x^l p_X(x) dX$ while first moment $\mu_x = m_1$ is called mean, other moments, $\sigma_x{}^2 = \mathbb{E}[X - \mu_x]^2$, $\Psi_3 = \mathbb{E}[X - \mu_x]^3/\sigma_x{}^3$, and $\Psi_4 = \mathbb{E}[X - \mu_x]^4/\sigma_x{}^4$ are called variance, coefficient of skewness and coefficient of kurtosis, respectively.

4.1.3 HILBERT SPACE

In a vector space of real random variables $X_1, X_2 \in \Omega(\Theta, \mathcal{F}, P)$, if it has a finite second moment $\left(\mathbb{E}[X_1]^2 < \infty, \mathbb{E}[X_2]^2 < \infty\right)$, the expectation operation allows the inner product and related norm to be expressed as

$$\langle X_1, X_2 \rangle = \mathbb{E}[X_1 X_2] \tag{4.1}$$

$$\|X\| = \sqrt{\mathbb{E}[X^2]} \tag{4.2}$$

Then it can be shown that $\Omega(\Theta, \mathcal{F}, P)$ is complete and form a Hilbert space [220].

DOI: 10.1201/9781003268048-4

4.2　RELIABILITY

4.2.1　SOURCES OF UNCERTAINTIES

In general, sources of uncertainties in engineering analysis problems are broadly divided into four; physical uncertainties, model uncertainties, human error, and estimation error. Physical uncertainties, also known as aleatory uncertainties, inherently exist in practical systems. These uncertainties are present in the system due to variability in the material properties, strength characteristics, geometry and loading conditions. On the other hand, assumptions and approximations cause model uncertainties. These uncertainties are also known as epistemic uncertainties. The sources of modeling error includes incorporated simplifications to achieve tractable model. Moreover, lack of proper understanding of the underlying physical phenomena also leads to error in the modeling and so the uncertainties. Nevertheless, these uncertainties reduced by acquiring the knowledge and understanding about the physics of the problem. Other than the aforementioned uncertainties, uncertainties due to the human errors are also common in engineering problems. Human errors represent the error involves in the stage of design, construction or operation. Finally, estimation errors which are statistical errors occur due to fluctuations in measurements, sampling and predictions.

4.2.2　RANDOM VARIABLES AND LIMIT STATE FUNCTION

As we stated earlier the variability of system parameters decides the reliability of the system. Here the first step to compute failure probability is defining the limit state function or performance criteria. The limit state function also represents the mathematical or functional relation of the parameters in the limiting state of the system.

Consider an N-dimensional vector of random variables $\boldsymbol{X}$, $\boldsymbol{X} = (X_1, X_2, ..., X_N) : \Omega_{\boldsymbol{X}} \to \mathbb{R}^N$, with probability density function $P_{\boldsymbol{X}}(\boldsymbol{x})$ and cumulative distribution function $F_{\boldsymbol{X}}(\boldsymbol{x}) = \mathbb{P}(\boldsymbol{X} \leq \boldsymbol{x})$, where $\mathbb{P}$ denotes the probability and $\Omega_{\boldsymbol{X}}$ denotes the probability space. Failure probability of a given system is quantified based on the limit state function $g(\boldsymbol{x}) = 0$. The function describes the reliability of the system such that $g(\boldsymbol{x}) < 0$ denotes the failure domain $\Omega_{\boldsymbol{X}}^F$

$$\Omega_{\boldsymbol{X}}^F \triangleq \{\boldsymbol{x} : g(\boldsymbol{x}) < 0\}, \tag{4.3}$$

and $g(\boldsymbol{x}) > 0$ denotes the safe region. $g(\boldsymbol{x}) = 0$ represents the limiting condition. The failure probability is defined as

$$P_f = \mathbb{P}(\boldsymbol{X} \in \Omega_{\boldsymbol{X}}^F) = \int_{\Omega_{\boldsymbol{X}}^F} dF_{\boldsymbol{X}}(\boldsymbol{x}) = \int_{\Omega_{\boldsymbol{X}}} I_{\Omega_{\boldsymbol{X}}^F} dF_{\boldsymbol{X}}(\boldsymbol{x}), \tag{4.4}$$

where $I_{\Omega_X^F}$ is a characteristic function and satisfies the following condition

$$I_{\Omega_X^F}(x) = \begin{cases} 1, & \text{if } x \in \Omega_X^F \\ 0, & \text{otherwise} \end{cases} \tag{4.5}$$

The vitality of the limit function $g(x)$ in determining the probability of failure is clearly vindicated. Here, the generalized multivariate form of limit state function can be represented as $y = g(x), g : \mathbb{R}^N \to \mathbb{R}$, with $N \geq 1$.

4.2.3 EARLIER METHODS

As we discussed previously, failure probability is obtained by evaluation of the joint probability over the failure domain. Nevertheless, in general, it is difficult to obtain the joint distribution of the random variables, even if it is available it will not be practically possible to evaluate the multi-variate integral.Therefore, a feasible approach to achieve the integral is utilization of analytical approximations, which simplifies the integral form. Most of the earlier structural reliability methods analyze the system performance by lumping the system variables into load (S) and resistance (R). Here the safety margin is expressed as

$$Z = R - S. \tag{4.6}$$

Further, a more generic expression in terms of Cornell's reliability index [51] can be expressed as

$$\beta_c = \frac{\mu_z}{\sigma_z}, \tag{4.7}$$

where μ_z is the mean and σ_z is the standard deviation of the variable z. When R and S are normally distributed the index, β_c can be expressed in terms of means (μ_S, μ_R) and standard deviations $(\sigma_S, \sigma_R$ of S, R and correlation coefficient $(\rho_R S)$ of the variable, and it is given by:

$$\beta_c = \frac{\mu_S - \mu_R}{\sqrt{\sigma_S^2 + \sigma_R^2 - 2\rho_{RS}\sigma_S\sigma_R}} \tag{4.8}$$

However, the above expression is not valid in a general case where the input of the limit state function is a N-dimensional random vector $X = X_1, X_2, X_3, ..., X_N{}^T$. Therefore, so-called mean value of first-order second moment reliability index is used, where Taylor series expansion truncated to linear terms is evaluated at the mean of the random vector μ_X. The index β_{MVFOSM} is given by:

$$\beta_{MVFOSM} = \frac{g(\mu_X)}{\sqrt{(\nabla g|_{X=\mu_x})^T R(\nabla g|_{X=\mu_x})}} \tag{4.9}$$

where, ∇ is $\{\partial/\partial X_1.....\partial/\partial X_N\}^T$.

4.3 SIMULATION METHODS IN UQ AND RELIABILITY

Simulation methods calculate the failure probability by employing sampling and estimation. The techniques work well with both implicit as well as explicit functions.

4.3.1 DIRECT MONTE CARLO SIMULATION

The most popular method for reliability analysis is perhaps the Direct Monte Carlo simulation (MCS) [163, 184]. MCS is a widely used method for computing multivariate integral in statistical physics. It performs simulations for the large number of sample points drawn independently from the probability distribution of the input variables. Therefore, in principle, the method can be employed to estimate failure probability by enumerating simulation results.

Consider the random vector $X = \{X_1, X_2, X3, ..., X_N\}$, joint probability $p_x(X)$ and limit state function $g(x)$, if m samples out of M samples generated from $p_x(X)$, $\{x_1, x_2, x_3, ..., x_M\}$ satisfy $g((x)) < 0$, then the estimation of failure is given by:

$$P_f \equiv P(g((x)) < 0) \cong \frac{m}{M} \tag{4.10}$$

Although the procedure is simple and straightforward, MCS becomes quite expensive as it has to utilize a large number of simulations to assure the convergence of a solution.

4.3.2 IMPORTANCE SAMPLING

To overcome the aforementioned problem, researchers have developed methods that enhance the computational efficiency of the crude MCS. Importance sampling is one among the methods [8, 67, 112]. The idea here is to concentrate the region of most importance in the distribution. For that the approach used is to generate samples from another sampling distribution which is expected to have a better approximation for the probability of failure from the new density function. Failure probability using the method is given by

$$P_F = \int_{g(x)<0} P_x(x)dx = \int_{g(x)<0} w(x)h(x) \tag{4.11}$$

where, $w(x) = p_x/h(x)$.

4.3.3 STRATIFIED SAMPLING

The domain of integration is separated into various parts in the stratified sampling approach [212], so that attention may be placed by simulating more from the regions that contribute to the failure event. If the whole domain of integration is partitioned into m mutually exclusive regions, then they are denoted by $R_1, R_2, R_3, ... R_m$. Using the total probability theorem, we can calculate the probability of of failure as

$$P_f = \sum_{j=1}^{m} (P(R_j)) \sum_{i=1}^{N_j} I_g(x_i) \qquad (4.12)$$

where, I is the indication function with respect to the performance function $g(\boldsymbol{x})$ and probabilities of failure $P(R_j)$ in the region R_j

$$I(\boldsymbol{x}) = \begin{cases} 1, & \text{if } \boldsymbol{x} \in \Omega_{\boldsymbol{X}}^F \\ 0, & \text{otherwise} \end{cases} \qquad (4.13)$$

Usually trial and error method is used as failure region is not known in advance.

4.3.4   DIRECTIONAL SAMPLING

In directional simulations [62] a reliability problem with a limit state/performance function $g(\boldsymbol{x})$ involving N normally distributed random variables is considered. The vector of random variable $\boldsymbol{X}$ can be written as product of length R and direction $\boldsymbol{A}$, where R^2 follows chi square distribution. The failure probability is evaluated by integrating conditional failure probability in the direction $\boldsymbol{A} = \boldsymbol{a}$.

$$\begin{aligned} P_F = P(g(x) \leq 0) &= P(g(RA) \leq 0) \\ &= \int\limits_{Alldirections} P(g(Ra) \leq 0) p_A(a) da \\ &= \int\limits_{Alldirections} (1 - \chi_N^2(r_a^2)) p_A(a) da \end{aligned} \qquad (4.14)$$

Directional simulations [62] have superior convergence rate as compared to crude MCS; however, these techniques still require a considerable number of simulations to achieve accurate estimations.

4.3.5   SUBSET SIMULATION

Another advanced simulation-based structural reliability approach is the subset simulation (SS) [9, 10, 15, 227]. This method is particularly suitable for cases having very small failure probability. The basic premise of SS is to represent the probability of failure as product of conditional probabilities which, in turn, are estimated by using the Metropolis Hasting algorithm. The literature of SS is quite mature and researchers over the year have come up with different variants. More recently, subset simulation has been combined with hybrid polynomial correlated function expansion [36] for estimating rare events in problems having expensive limit-state functions [37]. Here we briefly present the fundamentals of subset simulation.

Consider $\boldsymbol{X} = (X_1, X_2, \ldots, X_N) : \Omega_{\boldsymbol{X}} \to \mathbb{R}^N$ to be an N–dimensional random vector with cumulative distribution function $F_{\boldsymbol{X}}(\boldsymbol{x}) = \mathbb{P}(\boldsymbol{X} \leq \boldsymbol{x})$, where $\Omega_{\boldsymbol{X}}$ represents the probability space. We also consider $\Omega_{\boldsymbol{X}}^F$ to be the failure domain. With this setup, the probability of failure can be represented as:

$$
\begin{aligned}
P_f &= \mathbb{P}\left(\boldsymbol{X} \in \Omega_{\boldsymbol{X}}^F\right) \\
&= \int_{\Omega_{\boldsymbol{X}}^F} df_{\boldsymbol{X}}(\boldsymbol{x}) \\
&= \int_{\Omega_{\boldsymbol{X}}} \mathbb{I}_{\Omega_{\boldsymbol{X}}^F}(\boldsymbol{x})\, df_{\boldsymbol{X}}(\boldsymbol{x}).
\end{aligned}
\tag{4.15}
$$

$\mathbb{I}_k(\boldsymbol{x})$ in Equation 4.15 represents an indicator function such that

$$
\mathbb{I}_k(\boldsymbol{x}) = \begin{cases} 1 & \text{if } \boldsymbol{x} \in k \\ 0 & \text{elsewhere.} \end{cases}
\tag{4.16}
$$

Unfortunately, for most of the cases, the integral in Equation 4.15 is not solvable analytically and one has to resort to some numerical integration scheme. One such efficient method for solving the integration in Equation 4.15 is the SS. The primary idea of SS is to express the probability of failure as product of larger conditional probabilities. This algorithm was first proposed in [9] and since then, different variants of this algorithm has been proposed by different researchers [10, 15, 227]. Note that the SS algorithm generally works in the standard normal space. Therefore, using a mapping $\eta : \boldsymbol{x} \in \mathbb{R}^N \to \boldsymbol{z} \in \mathbb{R}^N$, we transform Equation 4.15 into the standard normal space

$$
P_f = \int_{\tilde{\Omega}_{\boldsymbol{Z}}} \mathbb{I}_{\tilde{\Omega}_{\boldsymbol{Z}}^F}(\boldsymbol{z})\, df_{\boldsymbol{Z}}(\boldsymbol{z}),
\tag{4.17}
$$

where $\boldsymbol{z}$ are realizations of the standard normal variables $\boldsymbol{Z}$. $\tilde{\Omega}_{\boldsymbol{Z}}$ and $\tilde{\Omega}_{\boldsymbol{Z}}^F$ are, respectively, the problem domain and failure domain in the standard normal space.

Typically, the SS algorithm starts at simulation level 0 where we generate N_s samples, $\boldsymbol{z}_i, i = 1, \ldots, N_s$ of $\boldsymbol{Z}$, transform $\boldsymbol{z}_i \, \forall i$ into $\boldsymbol{x}_i$ by using a mapping $\eta' : \boldsymbol{z} \in \mathbb{R}^N \to \boldsymbol{x} \in \mathbb{R}^N$ and then use $\boldsymbol{x}_i$ to evaluate the limit-state function $\mathcal{J}(\boldsymbol{x})$. Next, for simulation level 1, $p_0 N_s + 1$–th largest value is considered to be the threshold, where p_0 is the level probability chosen by the user. The $p_0 N_s$ samples are then used for generating additional samples conditioned on $\mathcal{J}(\boldsymbol{x}) > b_1$, where b_1 is the threshold for the current simulation level. The procedure continues until the level of interest is achieved. A step-by-step procedure for SS is shown in Algorithm 1.

Algorithm 1 Subset Simulation

Initialize: Provide statistical properties of the stochastic variables. Set p_0 and N_s.

$z_i \sim \mathcal{N}(0, \mathbf{I}_N)$ $i = 1, \ldots, N_s$, where $\mathbf{I}_N$ is a $N \times N$ identity matrix.

$x_i \leftarrow \eta'(z_i)$, $\forall i$.

$Y_i \leftarrow \mathcal{J}(x_i)$, $\forall i$.

Order the samples in $\{z_i^0, i = 1, \ldots, N_s\}$ in increasing order of the corresponding $Y_i : i = 1, \ldots, N_s$.

$c_0 \leftarrow p_0$ percentile of the samples $\{Y_i : i = 1, \ldots, N_s\}$. Set $F_1 = \{z \in \mathbb{R}^N : Y \leq c_0\}$.

Set $j = 0$.

repeat $c_j > 0$

$j \leftarrow j + 1$.

$l \leftarrow 1/p_0$

for $i = 1, \ldots, N_s p_0$ **do**

    for $k = 1, \ldots, l$ **do**

        Obtain new conditional samples for z_i and corresponding Y_i using Metropolis Hasting algorithm.

    end for

    Sort z according to Y.

    $c_j \leftarrow p_0$ percentile of the samples $\{Y_i : i = 1, \ldots, N_s\}$. Set $F_j = \{z \in \mathbb{R}^N : Y \leq c_0\}$.

end for

    Calculate the number of samples $N_f \left\{ z_k \in \tilde{\Omega}_z^F \right\}, k = 1, \ldots, N_f$.

$p_f \leftarrow p_0^{j-1} \frac{N_f}{N_s}$.

4.4 ROBUSTNESS

The overarching goal in engineering is to design a system/structure that is economical and safe. While this is straightforward in a deterministic settings, uncertainties in system often makes it challenging. Two separate designs under uncertainty algorithms, namely reliability-based design optimization (RBDO) and robust design optimization (RDO), are utilized. In RBDO, the goal is to ensure that the optimized system is reliable enough. This is achieved by including probabilistic constraints into the optimization framework. From a practical point-of-view, solution of RBDO is non-trivial as it involves a nested loop. While the outer loop solves the optimization problem, the inner loop solves a reliability analysis problem. Popular methods for solving RBDO problems include reliability index approach [66, 173], performance measure approach [186, 219], sequential optimization and reliability analysis approach [64], threshold shift method [86] and single loop approach [120]. Note that a number of studies used RBDO for designing various passive control devices [43, 131].

A viable alternative to the RBDO is the RDO. In RDO, the goal is to minimize the uncertainty propagated from the inputs to the response, i.e., the aim is to find insensitive design. This is achieved by solving a bi-objective optimization problem where we simultaneously minimize the mean and variance of the response. Consider x to be design variables, ξ to be the stochastic variables, and c to be the cost function, the objective function in RDO is represented as

$$c_{RTO} = \alpha\mu_c + (1 - \alpha)\sigma_c, \tag{4.18}$$

where μ_c represents the mean of the cost function and σ_c represents the standard deviation of the cost function. It is to be noted that in above expression, we have converted the bi-objective optimization problem into a single objective optimization problem by introducing a weight factor α where $0 \leq \alpha \leq 1$.

Similar to RBDO, solution of an RDO problem also involves a nested loop where the inner loop corresponds to the uncertainty quantification problem and the outer loop solves the optimization problem. As a consequence, RDO is computationally expensive. To address this issue, various approximation schemes have been developed by various researchers. This includes perturbation method [99, 213], point estimate method [94], polynomial chaos expansion (PCE) [171, 181, 214], Gaussian process (GP) [18, 19, 40], and radial basis function (RBF) [55, 111, 126]. Note that all these methods attempt to accelerate the solution procedure by solving the uncertainty quantification problem in an efficient manner. Unfortunately, the methods discussed above suffers from a number of issues. Perturbation method, for example, yield erroneous result if the underlying problem is highly nonlinear in nature. This is because perturbation method utilizes a second-order Taylor's series expansion. On the other hand, the surrogate-based approaches such as PCE, GP, and RBF suffer from the *curse of dimensionality*. To address these complications, adaptive [31, 159] and analytical [44] frameworks for solving RDO problems have also been proposed.

5 Digital Twin of Dynamic Systems

This chapter develops the digital twin model for a dynamic system. The background provided in the earliert chapters will now be used to develop a physics-based digital twin of a spring-mass system and a spring-mass-damper system. The influence of uncertainty on the system will be addressed.

5.1 DYNAMIC MODEL OF THE DIGITAL TWIN

In this section, the nominal dynamic system and the digital twin arising from this model are discussed. The nominal model is the "initial model" or the "starting model" of a digital twin. For engineering dynamic systems, the nominal model is a physics-based model which is verified, validated and calibrated. For example, this can be a finite element model of a ship or an aircraft when the product leaves the manufacturer facility and is ready to go into service. Its digital twin therefore begins the journey from the nominal model but as time passes, alters the original model such that it reflects the current state of the system. We explain the essential ideas of the digital twin through a single- degree-of-freedom (SDOF) dynamic system. The material presented in this chapter is adapted from Ref. [77].

5.1.1 SINGLE-DEGREE-OF-FREEDOM SYSTEM: THE NOMINAL MODEL

The equation of motion of a SDOF dynamic system [97] is given by

$$m_0 \frac{\mathrm{d}^2 u_0(t)}{\mathrm{d}t^2} + c_0 \frac{\mathrm{d}u_0(t)}{\mathrm{d}t} + k_0 u_0(t) = f_0(t) \tag{5.1}$$

This system given by Equation (5.1) is named as the nominal dynamic system. Here m_0, c_0, and k_0 are the nominal mass, damping and stiffness coefficients, respectively. The forcing function and the dynamic response is denoted by $f_0(t)$ and $u_0(t)$, respectively. The SDOF model in Equation (5.1) can represent a simplified model of a more complex dynamic system or can represent the dynamics in terms of modal coordinate of a multiple degree of freedom system. The formulations in this chapter is valid for either of these two important cases.

Dividing by m_0, the equation of motion (5.1) can be written as

$$\ddot{u}_0(t) + 2\zeta_0 \omega_0 \dot{u}_0(t) + \omega_0^2 u_0(t) = \frac{f(t)}{m_0} \tag{5.2}$$

DOI: 10.1201/9781003268048-5

The undamped natural frequency (ω_0) and the damping factor (ζ_0) are given as

$$\omega_0 = \sqrt{\frac{k_0}{m_0}} \tag{5.3}$$

$$\text{and}\quad \frac{c_0}{m_0} = 2\zeta_0\omega_0 \quad\text{or}\quad \zeta_0 = \frac{c_0}{2\sqrt{k_0 m_0}} \tag{5.4}$$

Th natural time period of the underlying undamped system is given by

$$T_0 = \frac{2\pi}{\omega_0} \tag{5.5}$$

Taking the Laplace transform of Equation (5.2) we get

$$s^2 U_0(s) + s2\zeta_0\omega_0 U_0(s) + \omega_0^2 U_0(s) = \frac{F_0(s)}{m_0} \tag{5.6}$$

where $U_0(s)$ and $F_0(s)$ are the Laplace transforms of $u_0(t)$ and $f_0(t)$, respectively. Solving the equation associated with coefficient of $U_0(s)$ in Equation (5.2) without the forcing term, yields the complex natural frequencies of the system

$$\lambda_{0_{1,2}} = -\zeta_0\omega_0 \pm i\omega_0\sqrt{1 - \zeta_0^2} = -\zeta_0\omega_0 \pm i\omega_{d_0} \tag{5.7}$$

Here the imaginary number $i = \sqrt{-1}$ and the damped natural frequency is expressed as

$$\omega_{d_0} = \omega_0\sqrt{1 - \zeta_0^2} \tag{5.8}$$

For a damped oscillator, at resonance, the frequency of oscillation is given by $\omega_{d_0} < \omega_0$. Therefore, for positive damping, the resonance frequency of a damped system is always lower than the corresponding underlying undamped system.

5.1.2 THE DIGITAL TWIN MODEL

Consider a physical system which can be well approximated by a single degree of freedom spring, mass and damper system as before.   Its digital twin equation can be written as

$$m(t_s)\frac{\partial^2 u(t, t_s)}{\partial t^2} + c(t_s)\frac{\partial u(t, t_s)}{\partial t} + k(t_s)u(t, t_s) = f(t, t_s) \tag{5.9}$$

Here t and t_s are the system time and a "slow time," respectively. The concept of "slow time" is important for understanding the evolution of digital twins. In contrast to the nominal system in Equation (5.1), $u(t, t_s)$ is a function of two variables and therefore the equation of motion is expressed in terms of the partial derivative with respect to the time variable t. The slow time or the service time t_s can be considered as a time variable which is much slower than

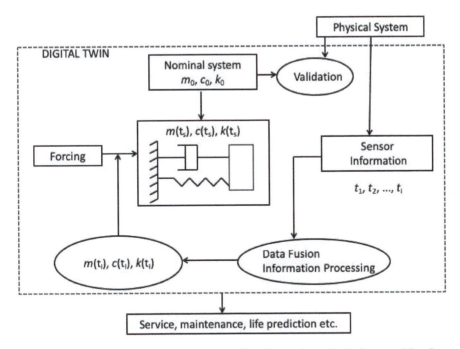

Figure 5.1 The overview of constructing a digital twin for a single-degree-of-freedom dynamic system.

t. For example, it could represent the number of cycles in an aircraft. Thus, mass $m(t_s)$, damping $c(t_s)$, stiffness $k(t_s)$ and forcing $F(t, t_s)$ change with t_s, for example due to system degradation during its service life. The forcing is also a function of time t and slow time t_s, as is the system response $x(t, t_s)$. Equation (5.9) is considered as a digital twin of the SDOF dynamic system. When $t_s = 0$, that is at the beginning of the service life of the system, the digital twin Equation (5.9) reduces to the nominal system in Equation (5.1).

It is assumed that sensors are placed on the physical system and take measurements at locations of time defined by t_s. The functional form of the relationship of mass, stiffness and forcing with t_s is unknown and has to be estimated from the measured sensor data. A general overview of how to create a digital twin for a SDOF dynamic system is shown in Figure 5.1.

Based on these discussions, the digital twin is defined as:

Definition. *The Digital Twin of a single-degree-of-freedom system is a bi-time-scale model reproducing the dynamics of the physical system at both time-scales.*

From Equation (5.9) it can be observed that the digital twin model is embodied in the functions $k(t_s)$, $m(t_s)$, and $c(t_s)$. Clearly, a wide variety of functional forms are possible. Considering the stiffness function as an example, $k(t_s)$ can be a deterministic or a random function (i.e., a random process). If $k(t_s)$ is a deterministic function, then the mathematical condition it must satisfy so that Equation (5.9) is solvable can be adumbrated in terms of a suitable norm. Therefore, the necessary conditions are

$$k(t_s) = k_0, \quad \text{when } t_s = 0$$

$$\text{and} \quad \int_0^{T_s} k^2(t_s) dt_s < \infty, \quad 0 < T_s < \infty \tag{5.10}$$

where T_s is the time till which the digital twin is to be constructed. If $k(t_s)$ is a random process, then its autocorrelation function must be finite at all time. Similar constraints are also imposed on $m(t_s)$ and $c(t_s)$.

A key characteristic of a digital twin is its connectivity. The recent development of the Internet of Things (IoT) presents numerous new data technologies and consequently drives the development of digital twin technology. IoT enables connectivity between the physical SDOF system and its digital counterpart. The basis of digital twins is based on this connection. Without connectivity, digital twin technology cannot exist; it is a product of the networked world. Connectivity is created by sensors on the physical system which gather data and integrate and communicate this data through various signal processing and communication technologies. The sensors sample data intermittently, typically, t_s represents discrete time points. It is assumed that changes in $k(t_s)$, $m(t_s)$ and $c(t_s)$ as so slow that the dynamics of system Equation (5.9) is effectively decoupled from these functional variations. Therefore, for all practical purposes $k(t_s)$, $m(t_s)$ and $c(t_s)$ are constant as far as the instantaneous dynamics of (5.9) is concerned as a function of t. We also assume that damping is small so that the effect of variations in $c(t_s)$ is negligible. Thus, only variations in the mass and stiffness are considered. The following functional forms are considered

$$k(t_s) = k_0(1 + \Delta_k(t_s))$$

$$\text{and} \quad m(t_s) = m_0(1 + \Delta_m(t_s)) \tag{5.11}$$

Based on the conditions in Equation (5.10), it follows that $\Delta_k(t_s) = \Delta_m(t_s) = 0$ for $t_s = 0$. In general, $k(t_s)$ is expected to be a decaying function over a long time used to represent a loss in the stiffness of the system. In an aircraft, this can represent stiffness loss due to matrix cracking, delamination and fiber breakage in the composite structure. Most aircraft are made from composite materials which are prone to these complex damage mechanisms. On the other hand, $m(t_s)$ can be an increasing or a decreasing function. For example, in the context of aircraft, it can represent the loading of cargo and passengers and also represent the use of fuel as the flight progresses.

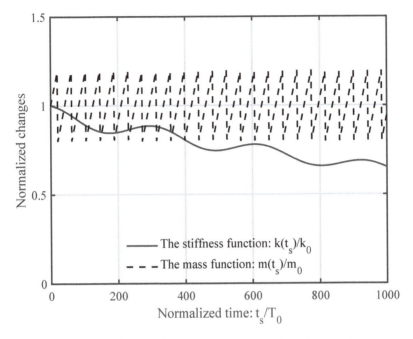

Figure 5.2 Examples of model functions representing long-term variabilities in the mass and stiffness properties of a digital twin system.

Based on the aforementioned constraints and discussions, the following representative functions are chosen as examples

$$\Delta_k(t_s) = e^{-\alpha_k t_s} \frac{(1 + \epsilon_k \cos(\beta_k t_s))}{(1 + \epsilon_k)} - 1 \qquad (5.12)$$

$$\text{and} \quad \Delta_m(t_s) = \epsilon_m \text{ SawTooth}(\beta_m(t_s - \pi/\beta_m)) \qquad (5.13)$$

Here SawTooth($\bullet$) represents a sawtooth wave with a period 2π. In Figure 5.2, overall variations in stiffness and mass properties arising from these function models are plotted as functions of time normalized to the natural time period of the nominal model. Numerical values used for these examples are $\alpha_k = 4 \times 10^{-4}$, $\epsilon_k = 0.05$, $\beta_k = 2 \times 10^{-2}$, $\beta_m = 0.15$ and $\epsilon_m = 0.25$. The rationale behind the selection of these functions is that the stiffness degrades over time in a periodic manner representing a possible fatigue crack growth in an aircraft over repeated pressurization; while the mass increases and decreases over the nominal value due to re-fuelling and fuel burn over a flight period. The key consideration is that a digital twin of the dynamical system such as an aircraft or other vehicle should be able to track these kinds of changes by exploiting sensor data measured on the system. Other systems would have different functional variations which can be obtained from domain experts in the area.

5.2 DIGITAL TWIN VIA STIFFNESS EVOLUTION

Advances in sensor technology are critical for the development of the digital twin concept. The formulation in this chapter uses the fact that natural frequency and response of the system can be measured online, as shown in some recent papers. Feng et al [71] developed a vision-based sensor system for remote measurement of structural displacements. Field tests carried out on a railway bridge and pedestrian bridge, demonstrated the accuracy of the sensor in both time and frequency domains. The vision-based non-contact sensor could extract structural displacement at any point on a structure. Wang et al [193] proposed a fiber Bragg grating sensor which uses vibration-induced strain to estimate the natural frequencies of the structure. Experiments were conducted using vibration and impact loads on metal pipes using an electrical strain gage, piezoelectric accelerometer and fiber Bragg grating sensor to get the natural frequencies. These studies demonstrate that natural frequency and displacement can be measured online by sensor systems, a fact which we will use in this chapter and in later chapters on the development of digital twins for dynamic systems.

5.2.1 EXACT NATURAL FREQUENCY DATA IS AVAILABLE

Consider that the mass and damping of the nominal model is unchanged and only the changes in the stiffness properties affect the digital twin. The equation of motion of the digital twin of a SDOF dynamic system with variation in the stiffness property only, for a fixed value of t_s is given by

$$m_0 \frac{\mathrm{d}^2 u(t)}{\mathrm{d}t^2} + c_0 \frac{\mathrm{d}u(t)}{\mathrm{d}t} + k_0(1 + \Delta_k(t_s))u(t) = f(t) \qquad (5.14)$$

This equation is a special case of the general equation in Equation (5.9). Since the value of t_s is fixed, the partial derivative terms in Equation (5.9) become total derivatives in this equation. Dividing with m_0 and solving the characteristic equation, the damped natural eigenfrequencies are expressed as

$$\lambda_{s_{1,2}}(t_s) = -\zeta_0 \omega_0 \pm i\omega_0 \sqrt{1 + \Delta_k(t_s) - \zeta_0^2} \qquad (5.15)$$

Here the subscript $(\bullet)_s$ denotes the "measured" values changing with the slow time t_s. Rearranging Equation (5.15) we obtain

$$\lambda_{s_{1,2}}(t_s) = -\underbrace{\frac{\zeta_0}{\sqrt{1 + \Delta_k(t_s)}}}_{\zeta_s(t_s)} \underbrace{\omega_0 \sqrt{1 + \Delta_k(t_s)}}_{\omega_s(t_s)}$$

$$\pm i \omega_0 \sqrt{1 + \Delta_k(t_s)} \underbrace{\sqrt{1 - \left(\frac{\zeta_0}{\sqrt{1 + \Delta_k(t_s)}}\right)^2}}_{\omega_{d_s}(t_s)} \qquad (5.16)$$

Here $\omega_s(t_s) = \omega_0 \sqrt{1 + \Delta_k(t_s)}$, $\zeta_s(t_s) = \zeta_0 / \sqrt{1 + \Delta_k(t_s)}$, and $\omega_{d_s}(t_s) = \omega_s(t_s) \sqrt{1 - \zeta_s^2(t_s)}$ are the natural frequency, damping factor and damped natural frequency of the digital twin, respectively. These three fundamental properties of system change with the slow time t_s.

For an SDOF model, several quantities can be measured and exploited to develop the digital twin. These quantities include but are not limited to transient response, forced response, frequency response, natural frequency and damping factor. The natural frequency measurement is one of the most fundamental quantities that can be obtained in a relatively simple manner. Typically, it is expected that sensor data from the system will be transmitted wirelessly to develop the digital twin. If the digital twin were to be constructed in real-time, it is imperative that the amount of data necessary is minimized. Considering these points, the choice of natural frequency as the "measured sensor data" is more effective than considering the time-domain response or the frequency domain response. This is mainly because the natural frequency is only a scalar for an SDOF system, while the response quantities can be vectors of inordinate length. It should be remembered that unlike the response measurements, the natural frequency is, typically, a derived quantity and not a direct measurement. However, several methods are available [71, 193] to extract frequency in real-time from measured vibration response data which can be implemented efficiently.

Most natural frequency extraction techniques obtain the damped natural frequency. Consider that this data is available for a given time instant $t_s \in [0, T_s]$ where T_s is the upper-bound of the time window over which the digital twin is to be created. Define the distance measure between two quantities A and B using the l_2 norm as

$$d(A, B) = \sqrt{(A - B)^H (A - B)} = \|A - B\|_2 \qquad (5.17)$$

Here $(\bullet)^H$ denotes the Hermitian transpose. For scalar quantities, $(\bullet)^H$ simply becomes the complex conjugate of $(\bullet)$. Considering only the damped natural frequency, the distance measure is expressed as

$$d_1(t_s) = d(\omega_{d_0}, \omega_{d_s}(t_s)) \qquad (5.18)$$

Using the corresponding expressions yields

$$\tilde{d}_1(t_s) = \frac{d_1(t_s)}{\omega_0} = \sqrt{1 - \zeta_0^2} - \sqrt{1 + \Delta_k(t_s) - \zeta_0^2} \qquad (5.19)$$

The quantity $\tilde{d}_1(t_s)$ is the distance measure normalized with respect to the undamped natural frequency of the nominal system. Here $\tilde{d}_1(t_s)$ is used as *the* input function for developing the digital twin from the original nominal model. In this case, the digital twin model is completely described by the function $\Delta_k(t_s)$. This must be calculated only from $\tilde{d}_1(t_s)$. This can be achieved by solving Equation (5.19) for $\Delta_k(t_s)$ as

$$\Delta_k(t_s) = -\tilde{d}_1(t_s)\left(2\sqrt{1 - \zeta_0^2} - \tilde{d}_1(t_s)\right) \qquad (5.20)$$

This is the exact solution for the digital twin and it is possible largely because an SDOF system is considered. For Multiple-degrees-of-freedom (MDOF) systems, a similar approach can be adopted. If the damping of the nominal system is small then $1 \gg \zeta_0^2$ which yields the approximate expression

$$\Delta_k(t_s) \approx -\tilde{d}_1(t_s)\left(2 - \tilde{d}_1(t_s)\right) \qquad (5.21)$$

If $\omega_{d_0} < \omega_{d_s}(t_s)$ then $\tilde{d}_1(t_s)$ will be positive. Consequently, $\Delta_k(t_s)$ from Equation (5.20) or Equation (5.21) will yield a negative value if $\tilde{d}_1(t_s) < 2, \forall t_s \in [0, T_s]$. This is what is physically expected. Recall that the function $\tilde{d}_1(t_s)$ is a dimensionless quantity denoting the relative shift in the natural frequency of the system as the system evolves in time. For most practical cases, $\tilde{d}_1(t_s) < 0.5, \forall t_s$ can be enforced. This is because if the natural frequency of a system changes by more than 50% over a period of time, then the physical validity of the original nominal model becomes questionable. In such situations, the original nominal model should be jettisoned and another model which represents the current physics accurately should be sought. In other words, it is inappropriate to create a digital twin based on the nominal model. As an absolute mathematical condition for the validity of Equations (5.20) and (5.21), $\tilde{d}_1(t_s)$ must satisfy

$$\tilde{d}_1(t_s) < 2, \forall t_s \qquad (5.22)$$

For all practical cases, $\tilde{d}_1(t_s)$ is expected to be significantly less than the above mathematical limit. We recommend about 15% of the above, that is, $\tilde{d}_1(t_s) < 0.3$, for the approach taken here to remain physically meaningful. These bounds and recommendations are given by us are based on our physical understanding of dynamic systems and may evolve in the future as the concept of digital twin becomes clearer and more case studies are reported.

5.2.2   NATURAL FREQUENCY DATA IS AVAILABLE WITH ERRORS

Data collected and transmitted by sensors onboard of a dynamic system (aerospace-vehicles, automobiles, wind turbines, ships, power plants, buildings, bridges, etc.) is susceptible to a multitude of errors. The error sources include measurement noise, transmission error, data storage inaccuracies, file corruptions, wireless signal loss, data bandwidth saturation, time-step mismatch, data hacking and alterations, to mention a few. Although the expression of $\Delta_k(t_s)$ to construct the digital twin is exact, the input data $d_1(t_s)$ typically contains error. This input data error percolates into the digital twin. This error is quantified following an approach similar to the previous subsection.

Suppose the error is denoted by $\theta(t_s)$. This can be a deterministic function (e.g., a bias in the sensors) or it can be a random function. For the second case, $\theta(t_s)$ becomes a random process [142] and must be defined by a suitable autocorrelation function. Considering this error, the measured damped natural frequency becomes

$$\widehat{\omega}_{d_s}(t_s) = \omega_0 \sqrt{1 + \Delta_k(t_s) - \zeta_0^2} + \theta(t_s) \qquad (5.23)$$

Combining this with Equations (5.18) and (5.19) yields

$$\widetilde{d}_1(t_s) = \frac{d_1(t_s)}{\omega_0} = \sqrt{1 - \zeta_0^2} - \sqrt{1 + \Delta_k(t_s) - \zeta_0^2} - \theta(t_s)/\omega_0 \qquad (5.24)$$

Solving this equation, the function $\Delta_k(t_s)$ to construct the digital twin is obtained as

$$\Delta_k(t_s) = -\left(\widetilde{d}_1(t_s) + \theta(t_s)/\omega_0\right)\left(2\sqrt{1 - \zeta_0^2} - \left(\widetilde{d}_1(t_s) + \theta(t_s)/\omega_0\right)\right) \quad (5.25)$$

Although this is the exact solution, the error in the measured data influences the digital twin. An alternative strategy is proposed next through which error in measured data is addressed in a more robust manner.

5.2.3   NATURAL FREQUENCY DATA IS AVAILABLE WITH ERROR ESTIMATES

To apply the approach in the previous subsection, a value of error in the measured data is necessary for all t_s. Unfortunately, this error value may not be always available or known precisely. A more plausible situation is when overall error estimate of the data is available. For example, this can be a manufacturer supplied tolerance of a sensing equipment (say 10%). In this case, without any loss of generality, we consider $\theta(t_s)$ is a zero-mean Gaussian white noise with a standard deviation σ_θ. Therefore

$$\mathrm{E}\left[\theta^2(t_s)\right] = \sigma_\theta^2 \quad \forall t_s \qquad (5.26)$$

where E [•] denotes the mathematical expectation operator. Taking the expectation of Equation (5.18), the distance measure is redefined as

$$\bar{d}_1^2(t_s) = \mathrm{E}\left[d^2\left(\omega_{d_0}, \widehat{\omega}_{d_s}(t_s)\right)\right] \tag{5.27}$$

Substituting the expression of $\widehat{\omega}_{d_s}(t_s)$ from Equation (5.23) in the above equation yields

$$\bar{d}_1^2(t_s) = \mathrm{E}\left[\left(\omega_0\sqrt{1-\zeta_0^2} - \omega_0\sqrt{1+\Delta_k(t_s)-\zeta_0^2} - \theta(t_s)\right)^2\right]$$

$$= \mathrm{E}\left[(d_1(t_s)-\theta(t_s))^2\right] = d_1^2(t_s) + \sigma_\theta^2 \tag{5.28}$$

Dividing this equation by ω_0^2, rearranging and taking the square root yields

$$\mathrm{sign}\left(\tilde{d}_1(t_s)\right)\sqrt{\tilde{d}_1^2(t_s) - \sigma_\theta^2/\omega_0^2} = \sqrt{1-\zeta_0^2} - \sqrt{1+\Delta_k(t_s)-\zeta_0^2} \tag{5.29}$$

Note that the sign of the square root is taken such that it keeps the sign of $\tilde{d}_1(t_s)$. Solving this equation, the function $\Delta_k(t_s)$ to develop the digital twin is

$$\Delta_k(t_s) = -\mathrm{sign}\left(\tilde{d}_1(t_s)\right)\sqrt{\tilde{d}_1^2(t_s) - \sigma_\theta^2/\omega_0^2}$$

$$\left(2\sqrt{1-\zeta_0^2} - \mathrm{sign}\left(\tilde{d}_1(t_s)\right)\sqrt{\tilde{d}_1^2(t_s) - \sigma_\theta^2/\omega_0^2}\right) \tag{5.30}$$

When the measurement becomes error free, that is, $\sigma_\theta^2 \to 0$, the above equation reduces to the deterministic case in Equation (5.20). Therefore, Equation (5.30) is the general expression of the function $\Delta_k(t_s)$ for the digital twin.

5.2.4 NUMERICAL ILLUSTRATIONS

To illustrate the applicability of the digital twin equations derived in the previous subsections, consider an SDOF system with nominal damping factor $\zeta_0 = 0.05$. The physical system evolves continuously in the "slow time" t_s. Consider that sensor data is transmitted intermittently with a certain regular time interval. For this example, the variation in the natural frequency is simulated using the change in the stiffness property of the system shown in Figure 5.2. In Figure 5.3(a) the actual change in the damped natural frequency of the system over time is shown along with sample discrete data points which are available for the digital twin.

Typically, the frequency of data availability depends on a number of practical details such as the bandwidth of the wireless data transmission system,

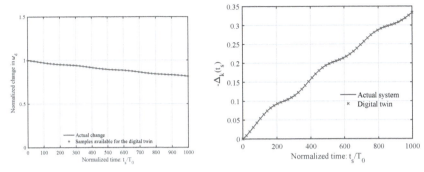

(a) Changes in (damped) natural frequency over time

(b) Digital twin constructed with the exact data

Figure 5.3 Changes in the (damped) natural frequency and the digital twin obtained using the exact data plotted as a function of the normalized "slow time" t_s/T_0. The digital twin in (b) is obtained from Equation (5.20) using the distance norm calculated from data in (a).

energy requirement of data collection, and cost of data transmission. The effectiveness of the digital twin depends on how frequently the data is sampled and transmitted and also how sharp is the variation of the measured frequency. For example, if a spike or unusual change in the frequency is missed due to poor sampling, the digital twin as proposed here will not be able to capture that real change in the system. Considering these limitations, Figure 5.3(b) shows the digital twin model obtained using Equation (5.20). Since Equation (5.20) is an exact equation, the digital twin exactly mirrors the actual system

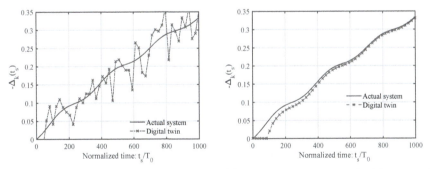

(a) Digital twin constructed with erroneous data

(b) Digital twin constructed with error estimate

Figure 5.4 Digital twin constructed with erroneous data and error estimates as a function of the normalized 'slow time' t_s/T_0 Error in the form of a discrete zero-mean Gaussian white noise with a standard deviation of 0.025 is considered.

for all values of t_s for which the natural frequency data is available. The validity of the digital twin given by Equation (5.20) is thus verified.

The case of erroneous data is shown in Figure 5.4. The digital twin in Figure 5.4(a) is obtained using Equation (5.25) and the one in Figure 5.4(b) is obtained using Equation (5.30). The error in the data is assumed to be a discrete zero-mean Gaussian white noise with standard deviation $\sigma_\theta = 0.025$. For the case when data error are incorporated in the digital twin, significant deviation from the actual system is observed in Figure 5.4(a). On the other hand, when the error estimate is known apriori, the digital twin tracks the real system closely as seen in Figure 5.4(b). Some discrepancies for lower values of t_s are present as in this region the value of the standard error is relatively more than that of the distance measure. This numerical example illustrates the significant impact of different types of error in the data and how they are processed on the effectiveness of the digital twin.

5.3 DIGITAL TWIN VIA MASS EVOLUTION

5.3.1 EXACT NATURAL FREQUENCY DATA IS AVAILABLE

Consider now that the stiffness and damping of the nominal model is unchanged and only the changes in the mass properties affect the digital twin. The equation of motion of the digital twin of a SDOF system for a fixed value of t_s, with variation only in the mass property is expressed as

$$m_0(1 + \Delta_m(t_s))\frac{\mathrm{d}^2 u(t)}{\mathrm{d}t^2} + c_0\frac{\mathrm{d}u(t)}{\mathrm{d}t} + k_0 u(t) = f(t) \tag{5.31}$$

Dividing with m_0 and solving the characteristic equation, the damped natural eigenfrequencies become

$$\lambda_{s_{1,2}}(t_s) = -\frac{\zeta_0\omega_0}{1 + \Delta_m(t_s)} \pm \mathrm{i}\frac{\omega_0\sqrt{1 + \Delta_m(t_s) - \zeta_0^2}}{1 + \Delta_m(t_s)} \tag{5.32}$$

Rearranging this equation yields

$$\lambda_{s_{1,2}}(t_s) = -\omega_s(t_s)\zeta_s(t_s) \pm \mathrm{i}\omega_{d_s}(t_s) \tag{5.33}$$

Here

$$\omega_s(t_s) = \omega_0/\sqrt{1 + \Delta_m(t_s)} \tag{5.34}$$

$$\zeta_s(t_s) = \zeta_0/\sqrt{1 + \Delta_m(t_s)} \tag{5.35}$$

$$\text{and} \quad \omega_{d_s}(t_s) = \omega_s(t_s)\sqrt{1 - \zeta_s^2(t_s)} = \frac{\omega_0\sqrt{1 + \Delta_m(t_s) - \zeta_0^2}}{1 + \Delta_m(t_s)} \tag{5.36}$$

are the natural frequency, damping factor and damped natural frequency of the digital twin, respectively. Again, these three fundamental properties of system change with the slow time t_s. Consider the distance measure similar to Equation (5.18) and define

$$d_2(t_s) = d(\omega_{d_0}, \omega_{d_s}(t_s)) \tag{5.37}$$

$$\text{or} \quad \tilde{d}_2(t_s) = \frac{d_2(t_s)}{\omega_0} = \sqrt{1 - \zeta_0^2} - \frac{\sqrt{1 + \Delta_m(t_s) - \zeta_0^2}}{(1 + \Delta_m(t_s))} \tag{5.38}$$

Here $\tilde{d}_2(t_s)$ is the distance measure normalized with respect to the undamped natural frequency of the nominal system. Consider $\tilde{d}_2(t_s)$ as *the* input function for developing the digital twin from the original nominal model. In this case, the digital twin model is completely described by the function $\Delta_m(t_s)$ and is obtained by solving Equation (5.38) for $\Delta_m(t_s)$ as

$$\Delta_m(t_s) = \frac{-2\tilde{d}_2(t_s)^2 + 4\tilde{d}_2(t_s)\sqrt{1 - \zeta_0^2} - 1 + 2\zeta_0^2}{+ \sqrt{1 - 4\tilde{d}_2(t_s)^2\zeta_0^2 + 8\tilde{d}_2(t_s)\sqrt{1 - \zeta_0^2}\zeta_0^2 - 4\zeta_0^2 + 4\zeta_0^4}}{2\left(-\tilde{d}_2(t_s) + \sqrt{1 - \zeta_0^2}\right)^2} \tag{5.39}$$

This is the exact solution; valid for any values of ζ_0. If a small damping assumption is made, then $\zeta_0^k \approx 0, k \geq 2$. Using this approximation, Equation (5.39) becomes

$$\Delta_m(t_s) \approx \frac{\tilde{d}_2(t_s)\left(2 - \tilde{d}_2(t_s)\right)}{\left(1 - \tilde{d}_2(t_s)\right)^2} \tag{5.40}$$

If $\tilde{d}_2(t_s)$ is positive, then $\Delta_m(t_s)$ must also be positive. A conclusion can be drawn from the above equation that the absolute mathematical condition for the validity of this analysis is that $\tilde{d}_2(t_s) < 2, \forall t_s$. However, it is recommended $\tilde{d}_2(t_s) < 0.3$ for the physical relevance of the original nominal model. The original nominal model should be examined carefully if $\tilde{d}_2(t_s)$ exceeds this value.

5.3.2 NATURAL FREQUENCY DATA IS AVAILABLE WITH ERRORS

Consider the error function $\theta(t_s)$; the measured damped natural frequency becomes

$$\hat{\omega}_{d_s}(t_s) = \frac{\omega_0\sqrt{1 + \Delta_m(t_s) - \zeta_0^2}}{1 + \Delta_m(t_s)} + \theta(t_s) \tag{5.41}$$

Combining this with Equations (5.37) and (5.38) yields

$$\tilde{d}_2(t_s) = \sqrt{1 - \zeta_0^2} - \frac{\sqrt{1 + \Delta_m(t_s) - \zeta_0^2}}{(1 + \Delta_m(t_s))} - \theta(t_s)/\omega_0 \tag{5.42}$$

Although the exact solution of this equation is possible, the resulting closed-form expression is cumbersome. Therefore, employing the small damping approximation, the function $\Delta_m(t_s)$ to construct the digital twin is obtained as

$$\Delta_m(t_s) \approx \frac{\left(\tilde{d}_2(t_s) + \theta(t_s)/\omega_0\right)\left(2 - \tilde{d}_2(t_s) - \theta(t_s)/\omega_0\right)}{\left(1 - \tilde{d}_2(t_s) - \theta(t_s)/\omega_0\right)^2} \tag{5.43}$$

For the validity of this expression $\tilde{d}_2(t_s) + \theta(t_s) < 2, \forall t_s$.

5.3.3 NATURAL FREQUENCY DATA IS AVAILABLE WITH ERROR ESTIMATES

Let $\theta(t_s)$ be a zero-mean Gaussian white noise with a standard deviation σ_θ. Taking the expectation of (5.37), define the distance measure as

$$\bar{d}_2^2(t_s) = \mathrm{E}\left[d^2\left(\omega_{d_0}, \hat{\omega}_{d_s}(t_s)\right)\right] \tag{5.44}$$

Substituting the expression of $\hat{\omega}_{d_s}(t_s)$ from Equation (5.41) in the above equation yields

$$\bar{d}_2^2(t_s) = \mathrm{E}\left[\left(\omega_0\sqrt{1-\zeta_0^2} - \frac{\omega_0\sqrt{1+\Delta_m(t_s)-\zeta_0^2}}{1+\Delta_m(t_s)} - \theta(t_s)\right)^2\right] \tag{5.45}$$

$$= \mathrm{E}\left[\left(d_2(t_s) - \theta(t_s)\right)^2\right] = d_2^2(t_s) + \sigma_\theta^2$$

Dividing this equation by ω_0^2, rearranging and taking the square root gives

$$\mathrm{sign}\left(\tilde{d}_2(t_s)\right)\sqrt{\bar{d}_2^2(t_s) - \sigma_\theta^2/\omega_0^2} = \sqrt{1-\zeta_0^2} - \frac{\sqrt{1+\Delta_m(t_s)-\zeta_0^2}}{(1+\Delta_m(t_s))} \tag{5.46}$$

The sign of the square root is taken to keep the sign of $\tilde{d}_1(t_s)$. Solving this equation with the small damping approximation, the function $\Delta_m(t_s)$ to construct the digital twin becomes

$$\Delta_m(t_s) \approx \frac{\mathrm{sign}\left(\tilde{d}_2(t_s)\right)\sqrt{\bar{d}_2^2(t_s) - \sigma_\theta^2/\omega_0^2}\left(2 - \mathrm{sign}\left(\tilde{d}_2(t_s)\right)\sqrt{\bar{d}_2^2(t_s) - \sigma_\theta^2/\omega_0^2}\right)}{\left(1 - \mathrm{sign}\left(\tilde{d}_2(t_s)\right)\sqrt{\bar{d}_2^2(t_s) - \sigma_\theta^2/\omega_0^2}\right)^2} \tag{5.47}$$

When the measurement becomes error free, that is, $\sigma_\theta^2 \to 0$, the above equation reduces to the deterministic case in Equation (5.40). Therefore, Equation (5.47) can be considered as the general expression of the function $\Delta_m(t_s)$ for the digital twin.

5.3.4 NUMERICAL ILLUSTRATIONS

To illustrate the applicability of the digital twin equations derived for mass evolution, consider an SDOF system with nominal damping factor $\zeta_0 = 0.05$.

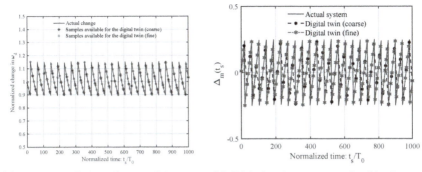

(a) Changes in (damped) natural frequency (b) Digital twin constructed with the exact
over time   data

Figure 5.5   Changes in the (damped) natural frequency and the digital twin obtained
using the exact data plotted as a function of the normalized "slow time" t_s/T_0. The
effect of two different sampling rates is shown. The digital twin in (b) is obtained
from Equation (5.20) using the distance norm calculated from data in (a).

The physical system evolves continuously in the slow time t_s. For numerical
illustrations, it is considered that the sensor data is available at two different
sampling rate. The variation in the natural frequency is simulated using the
change in the mass property of the system shown in Figure 5.2. In Figure
5.5(a) the actual change in the damped natural frequency of the system is
shown over the time along with a coarse and a fine sample of discrete data
points which is available for the digital twin. The coarse samples of data omit
certain features of the underlying variations, highlighting the role of sampling
rate on the digital twin. In Figure 5.5(b), the digital twin model obtained
using Equation (5.40) is shown. As Equation (5.40) is an exact equation, the
digital twin perfectly represents the actual system for all values of t_s for which
the natural frequency data is available. This verifies the veracity of the digital
twin given by Equation (5.40).

The case of erroneous data is shown in Figure 5.6. The digital twin in
Figure 5.6(a) is obtained using (5.43). The digital twin in Figure 5.6(b) is
obtained using (5.47). The error in the data is modeled as a discrete zero-
mean Gaussian white noise with standard deviation $\sigma_\theta = 0.025$. The case
where data errors are incorporated in the digital twin leads to deviation from
the actual system as observed in Figure 5.6(a). The deviations become more
pronounced when data sampling is coarse.

On the other hand, if the error estimate is apriori known, the digital twin
follows the real system more closely as shown in Figure 5.6(b). There are
some discrepancies due to coarse data sampling as expected. This numerical
example highlights the significant impact of different types of error in the data
on the effectiveness of the digital twin.

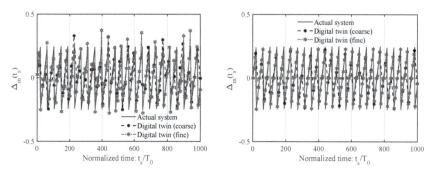

(a) Digital twin constructed with erroneous data

(b) Digital twin constructed with error estimate

Figure 5.6 Digital twin constructed with erroneous data and error estimates as a function of the normalized "slow time" t_s/T_0. Error in the form of a discrete zero-mean Gaussian white noise with a standard deviation of 0.025 is considered.

5.4 DIGITAL TWIN VIA MASS AND STIFFNESS EVOLUTION

Now consider that both the stiffness and mass properties of the system simultaneously change with the slow time-scale t_s. The damping of the nominal model is unchanged. The equation of motion of the digital twin of a SDOF system for a fixed value of t_s with variations in the stiffness and mass properties is given by

$$m_0(1 + \Delta_m(t_s))\frac{d^2 u(t)}{dt^2} + c_0\frac{du(t)}{dt} + k_0(1 + \Delta_k(t_s))u(t) = f(t) \quad (5.48)$$

Dividing with m_0 and solving the characteristic equation yields the damped natural eigenfrequencies

$$\lambda_{s_{1,2}}(t_s) = -\frac{\zeta_0 \omega_0}{1 + \Delta_m(t_s)} \pm i\frac{\omega_0 \sqrt{(1 + \Delta_k(t_s))(1 + \Delta_m(t_s)) - \zeta_0^2}}{1 + \Delta_m(t_s)} \quad (5.49)$$

Rearranging this equation gives

$$\lambda_{s_{1,2}}(t_s) = -\omega_s(t_s)\zeta_s(t_s) \pm i\omega_{d_s}(t_s) \quad (5.50)$$

Here

$$\omega_s(t_s) = \omega_0 \frac{\sqrt{1 + \Delta_k(t_s)}}{\sqrt{1 + \Delta_m(t_s)}}, \quad (5.51)$$

$$\zeta_s(t_s) = \frac{\zeta_0}{\sqrt{1 + \Delta_k(t_s)}\sqrt{1 + \Delta_m(t_s)}}, \quad (5.52)$$

$$\text{and} \quad \omega_{d_s}(t_s) = \omega_s(t_s)\sqrt{1 - \zeta_s^2(t_s)} = \frac{\omega_0\sqrt{(1 + \Delta_k(t_s))(1 + \Delta_m(t_s)) - \zeta_0^2}}{1 + \Delta_m(t_s)} \quad (5.53)$$

are the natural frequency, damping factor, and damped natural frequency of the digital twin, respectively. As mentioned before, these three fundamental properties of system change with the slow time t_s.

Unlike the previous two cases considered, here we have two unknown functions, namely, $\Delta_k(t_s)$ and $\Delta_m(t_s)$, which define the digital twin. To obtain these two unknown functions uniquely, two independent equations are necessary for all t_s. From Equations (5.51)–(5.53) observe that all the three key dynamic quantities, namely, the natural frequency, damping factor, and damped natural frequency change as functions of both $\Delta_k(t_s)$ and $\Delta_m(t_s)$. Therefore, it is sufficient to consider two out of these three dynamic quantities to measure for establishing the digital twin. The choice of which two quantities should be considered depends on what data is available as neither of these quantities are "direct measurements" for a real-life dynamic system. In the following, we again consider three physically realistic cases through which a digital twin can be established.

5.4.1   EXACT NATURAL FREQUENCY DATA IS AVAILABLE

The damped natural eigenfrequencies give in Equation (5.49) are complex valued quantities. Now consider the real and imaginary parts separately to develop two equations through which the digital twin may be established. Considering the distance measure similar to (5.18) and applying it separately to the real and imaginary parts we define

$$d_\Re(t_s) = d(\Re(\lambda_0), \Re(\lambda_s(t_s))) \tag{5.54}$$

$$\text{and} \quad d_\Im(t_s) = d(\Im(\lambda_0), \Im(\lambda_s(t_s))) \tag{5.55}$$

Dividing the distance measures by ω_0 the normalized error measures are given as

$$\widetilde{d}_\Re(t_s) = \frac{d_\Re(t_s)}{\omega_0} = \frac{\zeta_0}{1 + \Delta_m(t_s)} - \zeta_0 \tag{5.56}$$

$$\text{and} \quad \widetilde{d}_\Im(t_s) = \frac{d_\Im(t_s)}{\omega_0} = \sqrt{1 - \zeta_0^2} - \frac{\sqrt{(1 + \Delta_k(t_s))(1 + \Delta_m(t_s))} - \zeta_0^2}{1 + \Delta_m(t_s)} \tag{5.57}$$

We use $\widetilde{d}_\Re(t_s)$ and $\widetilde{d}_\Im(t_s)$ as the input functions for developing the digital twin from the original nominal model. In this case, the digital twin model is completely defined by the functions $\Delta_k(t_s)$ and $\Delta_m(t_s)$ and can be obtained by solving the above two equations simultaneously as

$$\Delta_m(t_s) = -\frac{\widetilde{d}_{\Re}(t_s)}{\zeta_0 + \widetilde{d}_{\Re}(t_s)} \tag{5.58}$$

and

$$\Delta_k(t_s) = \frac{\zeta_0\,\widetilde{d}_{\Re}^2(t_s) - \left(1 - 2\,\zeta_0{}^2\right)\widetilde{d}_{\Re}(t_s) - 2\,\sqrt{1 - \zeta_0{}^2}\,\zeta_0\,\widetilde{d}_{\Im}(t_s) + \zeta_0\,\widetilde{d}_{\Im}^2(t_s)}{\zeta_0 + \widetilde{d}_{\Re}(t_s)} \tag{5.59}$$

This is the exact solution and it is valid for any values of ζ_0. It is evident that to establish a digital twin with simultaneous mass and stiffness evolutions, we need distance measures for both real and imaginary parts of the complex natural frequency.

5.4.2 EXACT NATURAL FREQUENCY DATA IS AVAILABLE WITH ERRORS

The real and imaginary parts of a complex natural frequency physically correspond to significantly different aspects of the system dynamics even they are part of a single eigenvalue. The real part corresponds to the decay rate of the free vibration response. The imaginary part corresponds to the oscillation frequency. The techniques to measure these quantities [70] also differ considerably. Therefore, the error corresponding to these two quantities are in general different and independent to each other.

Consider the error functions $\theta_{\Re}(t_s)$ and $\theta_{\Im}(t_s)$. The measured complex eigenfrequency becomes

$$\widehat{\lambda}_{s_{1,2}}(t_s) = -\left(\frac{\zeta_0\omega_0}{1 + \Delta_m(t_s)} + \theta_{\Re}(t_s)\right)$$
$$\pm \mathrm{i}\left(\frac{\omega_0\sqrt{(1 + \Delta_k(t_s))\,(1 + \Delta_m(t_s))} - \zeta_0^2}{1 + \Delta_m(t_s)} + \theta_{\Im}(t_s)\right) \tag{5.60}$$

Combining this with Equations (5.54)–(5.57) we have

$$\widetilde{d}_{\Re}(t_s) = \frac{\zeta_0}{1 + \Delta_m(t_s)} - \zeta_0 + \theta_{\Re}(t_s)/\omega_0 \tag{5.61}$$

$$\text{and} \quad \widetilde{d}_{\Im}(t_s) = \sqrt{1 - \zeta_0{}^2} - \frac{\sqrt{(1 + \Delta_k(t_s))\,(1 + \Delta_m(t_s))} - \zeta_0^2}{1 + \Delta_m(t_s)} - \theta_{\Im}(t_s)/\omega_0 \tag{5.62}$$

Solving the above two equations simultaneously, functions $\Delta_k(t_s)$ and $\Delta_m(t_s)$ can be derived to describe the digital twin model with error in the data as

$$\Delta_m(t_s) = -\frac{\widetilde{d}_{\Re}(t_s) - \theta_{\Re}(t_s)/\omega_0}{\zeta_0 + \widetilde{d}_{\Re}(t_s) - \theta_{\Re}(t_s)/\omega_0} \tag{5.63}$$

and

$$\zeta_0 \left(\widetilde{d}_{\Re}(t_s) - \theta_{\Re}(t_s)/\omega_0 \right)^2 - \left(1 - 2\zeta_0^2 \right) \left(\widetilde{d}_{\Re}(t_s) - \theta_{\Re}(t_s)/\omega_0 \right)$$

$$\Delta_k(t_s) = \frac{- 2\sqrt{1 - \zeta_0^2}\zeta_0 \left(\widetilde{d}_{\Im}(t_s) + \theta_{\Im}(t_s)/\omega_0 \right) + \zeta_0 \left(\widetilde{d}_{\Im}(t_s) + \theta_{\Im}(t_s)/\omega_0 \right)^2}{\zeta_0 + \widetilde{d}_{\Re}(t_s) - \theta_{\Re}(t_s)/\omega_0}$$

(5.64)

The error functions $\theta_{\Re}(t_s)$ and $\theta_{\Im}(t_s)$ can be deterministic or random functions. If they are random functions, then the complete description should be given cross correlation function of these two random processes. In this chapter, we assume them to be statistically uncorrelated functions.

5.4.3 EXACT NATURAL FREQUENCY DATA IS AVAILABLE WITH ERROR ESTIMATES

Proceeding further from the discussion in the previous subsection, we consider $\theta_{\Re}(t_s)$ and $\theta_{\Im}(t_s)$ are statistically uncorrelated zero-mean Gaussian white noises with a standard deviations $\sigma_{\theta_{\Re}}$ and $\sigma_{\theta_{\Im}}$. Take the expectation of (5.54) and (5.55), and define new distance measures as

$$\bar{d}_{\Re}^2(t_s) = \mathrm{E}\left[d^2(\Re(\lambda_0), \Re(\widehat{\lambda}_s(t_s))) \right]$$

(5.65)

$$\text{and} \quad \bar{d}_{\Im}^2(t_s) = \mathrm{E}\left[d^2(\Im(\lambda_0), \Im(\widehat{\lambda}_s(t_s))) \right]$$

(5.66)

Substituting the expression of $\widehat{\lambda}_s(t_s)$ from Equation (5.60) in the above equation and following the procedure outlined in the previous sections yields

$$\mathrm{sign}\left(\widetilde{d}_{\Re}(t_s) \right) \sqrt{\widetilde{d}_{\Re}^2(t_s) - \sigma_{\theta_{\Re}}^2/\omega_0^2} = \frac{\zeta_0}{1 + \Delta_m(t_s)} - \zeta_0$$

(5.67)

$$\text{and} \quad \mathrm{sign}\left(\widetilde{d}_{\Im}(t_s) \right) \sqrt{\widetilde{d}_{\Im}^2(t_s) - \sigma_{\theta_{\Im}}^2/\omega_0^2} = \sqrt{1 - \zeta_0^2}$$

$$- \frac{\sqrt{(1 + \Delta_k(t_s))(1 + \Delta_m(t_s))} - \zeta_0^2}{1 + \Delta_m(t_s)}$$

(5.68)

Solving these two coupled equations, the functions $\Delta_k(t_s)$ and $\Delta_m(t_s)$ become

$$\Delta_m(t_s) = -\frac{\mathrm{sign}\left(\widetilde{d}_{\Re}(t_s) \right) \sqrt{\widetilde{d}_{\Re}^2(t_s) - \sigma_{\theta_{\Re}}^2/\omega_0^2}}{\zeta_0 + \mathrm{sign}\left(\widetilde{d}_{\Re}(t_s) \right) \sqrt{\widetilde{d}_{\Re}^2(t_s) - \sigma_{\theta_{\Re}}^2/\omega_0^2}}$$

(5.69)

$$\text{and } \Delta_k(t_s) = \frac{\zeta_0 \left(\widetilde{d}_{\Re}^2(t_s) - \sigma_{\theta_{\Re}}^2/\omega_0^2 \right) - \left(1 - 2\zeta_0^2 \right) \mathrm{sign}\left(\widetilde{d}_{\Re}(t_s) \right) \sqrt{\widetilde{d}_{\Re}^2(t_s) - \sigma_{\theta_{\Re}}^2/\omega_0^2}}{\begin{array}{c} -2\sqrt{1 - \zeta_0^2}\zeta_0 \,\mathrm{sign}\left(\widetilde{d}_{\Im}(t_s) \right) \sqrt{\widetilde{d}_{\Im}^2(t_s) - \sigma_{\theta_{\Im}}^2/\omega_0^2} + \zeta_0 \left(\widetilde{d}_{\Im}^2(t_s) - \sigma_{\theta_{\Im}}^2/\omega_0^2 \right) \\ \hline \zeta_0 + \mathrm{sign}\left(\widetilde{d}_{\Re}(t_s) \right) \sqrt{\widetilde{d}_{\Re}^2(t_s) - \sigma_{\theta_{\Re}}^2/\omega_0^2} \end{array}}$$

(5.70)

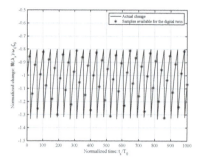 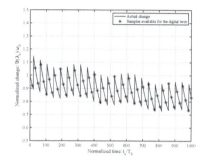

(a) Changes in the real part of the natural frequency

(b) Chnges in the imaginary part of the natural frequency

Figure 5.7 Normalized changes in the real and imaginary parts of the natural frequency as a function of the normalized "slow time" t_s/T_0. The reduced number of samples available to construct the digital twin are are marked by a '*'.

This is the exact solution and it is valid for any values of ζ_0. When the measurement becomes bereft of error, that is, $\sigma^2_{\theta_\Re}, \sigma^2_{\theta_\Im} \to 0$, the above equations yield the deterministic case in Equations (5.58) and (5.59).

5.4.4 NUMERICAL ILLUSTRATIONS

We again employ the previously used SDOF model to gain an understanding of the digital twin equations derived for simultaneous mass and stiffness evolution. The variation in the natural frequency as a function of the slow time t_s is simulated using the change in the mass and stiffness properties of the system as is shown in Figure 5.2. In Figure 5.7, the actual change in the real and imaginary parts of the natural frequency of the system over time are shown. The reduced number of samples available to construct the digital twin is also provided in the figure. The damping factor is considered to be $\zeta_0 = 0.05$. Unlike for the previous two cases, the presence of damping is now crucial to construct the digital twin using the mass and stiffness evolution. In Figure 5.8, we illustrate the actual simultaneous changes in the mass and stiffness properties and the values obtained using the digital twin Equations (5.58) and (5.59). As the data is coarse, there is a need to interpolate the digital twin between the data points. The interpolation of the stiffness function resembles the actual system well as the functional variation is smooth and mathematically continuous in nature. However, the mass function is not interpolated with the same accuracy as the variations are sharp and not mathematically continuous in nature. This illustrates that different aspects of the same digital twin can be affected in a very different manner under the same data-resolution.

Now consider the effect of random errors in the data on the digital twin. In Figure 5.9 the mass and stiffness properties using the digital twin Equations (5.63) and (5.64) are shown. These results consider the standard deviation of

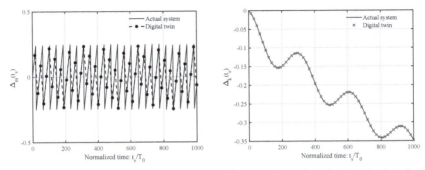

(a) The mass function for the digital twin   (b) The stiffness function for the digital twin

Figure 5.8 Digital twin obtained from coarse but exact data via simultaneous mass and stiffness evolution as a function of the normalized "slow time" t_s/T_0. The mass function in (a) is obtained using Equation (5.58) while the stiffness function in (b) is obtained using Equation (5.59).

the errors in the imaginary and real parts of the complex natural frequencies are 0.025 and $0.025\zeta_0$, respectively. The impact of the error in the data is most severe on the stiffness evolution function. The mass evolution function of the digital twin, on the other hand, is less affected by the error in the data. In Figure 5.10, the digital twin obtained with error estimates is shown as a function of the normalized "slow time" t_s/T_0. The evolution of the mass

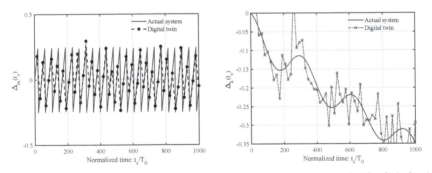

(a) The mass function for the digital twin   (b) The stiffness function for the digital twin

Figure 5.9 Digital twin obtained from erroneous data via simultaneous mass and stiffness evolution as a function of the normalized "slow time" t_s/T_0. The mass function in (a) is obtained using Equation (5.63) while the stiffness function in (b) is obtained using Equation (5.64). Errors in the form of a discrete zero-mean Gaussian white noise with standard deviations of 0.025 and $0.025\zeta_0$ are considered for the imaginary and real parts.

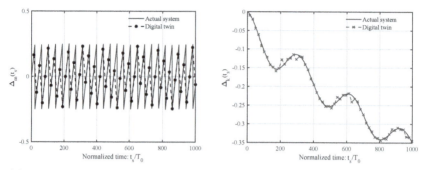

(a) The mass function for the digital twin   (b) The stiffness function for the digital twin

Figure 5.10   Digital twin obtained with error estimates via simultaneous mass and stiffness evolution as a function of the normalized "slow time" t_s/T_0. The mass function in (a) is obtained using Equation (5.69) while the stiffness function in (b) is obtained using Equation (5.70). Errors in the form of a discrete zero-mean Gaussian white noise with standard deviations of 0.025 and $0.025\zeta_0$ are considered for the imaginary and real parts.

and stiffness properties are obtained from Equations (5.69) and (5.70). The standard deviation of the errors in the imaginary and real parts of the complex natural frequencies are assumed as before. Figure 5.10(a) shows that the mass evolution function of the digital twin is not significantly perturbed by the error in the data compared to what noticed in Figure 5.9(a). On the other hand, a significant difference is observed for the stiffness evolution function between what is obtained in Figure 5.9(b) and Figure 5.10(b). It is clear that if an error estimate of the data is available, the application of Equation (5.70) leads to a realistic digital twin.

5.5   DISCUSSIONS

A digital twin of physical systems can be achieved in various ways. The majority of existing works focus on broader conceptual aspects of a digital twin. The aim of this chapter is to consider the specific case of structural dynamic systems. In particular, a SDOF system is considered. Key ideas introduced in this chapter include:

> The equation of motion is expressed in terms of two time variables which are independent of each other. Time t denotes the (fast) time describing the system dynamics. Time t_s denotes a (slow) time describing the evolution of the digital twin. This separation of the time-scale is crucial for the practical development of digital twins of engineering dynamic systems.
>
> There must be a nominal model of the system from which a digital twin evolves. The nominal model is a validated, calibrated and verified

the model of the system at time $t_s = 0$. The nominal model makes use of the considerable expertise developed in the area of modeling of systems.

Variations in the key response descriptors (natural frequency and damping factor in this case) should not deviate from the nominal system by more than 25%.

Noise in the sensor data can be absorbed in two different ways—either directly or by statistical processes.

Exact and approximate closed-form mathematical expressions to explicitly obtain the digital twin of an SDOF dynamical system under different physically realistic contexts are derived.

Model functions describing stiffness and mass variations, coarse sampling and associated uncertainty models are introduced in the numerical illustrations.

The system studied here is a simple SDOF dynamic system. This is expressed by a second-order ordinary differential equation. The closed-form digital twin expressions are directly applicable to other physical problems (e.g., electrical circuits) governed by this kind of equations. Although the work presented here considers only a SDOF system, the underlying conceptual framework can form the basis for more rigorous theoretical investigations encompassing a wider variety of practical problems, some of which will be discussed in later chapters. Some possibilities include investigating the following urgent issues:

1. *Multi time-scale digital twins*: This chapter introduces the idea of one time-scale for the evolution of the digital twin. However, there is no physical or mathematical reason as to why this must be restricted to only one time scale. It is possible that factors in a complex digital twin evolve at different time scales. For example, the mass of a system can change due to corrosion, while the stiffness of a system can degrade due to fatigue. These two processes have different time scale of evolution. Therefore, in a more general setting, the equation of motion of a dynamic digital twin become

$$m(t_{s_1}, t_{s_2}, t_{s_3}, \cdots) \frac{\partial^2 u(t, t_{s_1}, t_{s_2}, t_{s_3}, \cdots)}{\partial t^2}$$
$$+ c(t_{s_1}, t_{s_2}, t_{s_3}, \cdots) \frac{\partial u(t, t_{s_1}, t_{s_2}, t_{s_3}, \cdots)}{\partial t}$$
$$+ k(t_{s_1}, t_{s_2}, t_{s_3}, \cdots) u(t, (t_{s_1}, t_{s_2}, t_{s_3}, \cdots) = f(t, t_{s_1}, t_{s_2}, t_{s_3}, \cdots) \quad (5.71)$$

Here $u(t, t_{s_1}, t_{s_2}, t_{s_3}, \cdots)$ is a multivariate function of not only the system time t but also the independent multiple slow times $t_{s_1}, t_{s_2}, t_{s_3}, \cdots$.

2. *Digital twins from the response in the time or frequency domain*: Natural frequency and damping factor measurements are used in this chapter for establishing the digital twin. These are discrete scalar numbers which are normally derived/estimated from response measurements in the time

or frequency domain. Methods can be developed which will establish a digital twin directly from the response measurements. This is useful as real dynamic systems may not have onboard capabilities for reliable extraction of natural frequencies and damping factors. In other words, digital twin should be developed by using continuous real-time dynamic measurements.

3. *Multiple-degrees-of-freedom (MDOF) digital twins*: The SDOF model is a simple idealization of complex MDOF systems. For an effective digital twin with realistic predictive capabilities, damped MDOF models (see for example [2, 97]) are important. The equation of motion of a MDOF digital twin can be expressed by

$$\mathbf{M}(t_s)\frac{\partial^2\mathbf{u}(t,t_s)}{\partial t^2} + \mathbf{C}(t_s)\frac{\partial\mathbf{u}(t,t_s)}{\partial t} + \mathbf{K}(t_s)\mathbf{u}(t,t_s) = \mathbf{f}(t,t_s) \qquad (5.72)$$

Here $\mathbf{M}(t_s)$, $\mathbf{C}(t_s)$ and $\mathbf{K}(t_s)$ are $N \times N$ matrices and $\mathbf{u}(t,t_s)$ and $\mathbf{f}(t,t_s)$ are N dimensional vectors. A set of eigenvalues and corresponding eigenvectors can be utilized to construct the digital twin model.

4. *Stochastically parametrized digital twins*: The mass $m(t_s)$, damping $c(t_s)$, a stiffness $k(t_s)$ functions are assumed to be deterministic. However, it was established that there are uncertainties in the measured and transmitted data available to construct the digital twin. In this chapter, the effect of these uncertainties has been ameliorated by using statistical averages. While this can be considered as a legitimate first step, a more rigorous approach is to model $m(t_s)$, $c(t_s)$ and $k(t_s)$ themselves as random quantities. As they are functions of the time variable t_s, each of these functions must be modeled as a random process. A stationary Gaussian random process model with an assumed autocorrelation function can be used for modeling these functions. The establishment of the digital twin will be through the estimation of the parameters of the autocorrelation function.

5. *Nonlinear digital twins*: The evolution of the mass $m(t_s)$, damping $c(t_s)$ and stiffness $k(t_s)$ functions are considered as nonlinear functions of t_s in this chapter. However, the dynamics of the system in t is considered as linear. For many digital twins, a linear dynamic assumption may not be physically accurate. One example is dynamic systems with large-amplitude vibrations. Several types of nonlinearities can be considered for the digital twin. An illustration of a cubic-type nonlinearity (see for example [137]) can be achieved using a Duffing's oscillator digital twin as

$$m(t_s)\frac{\partial^2 u(t,t_s)}{\partial t^2} + c(t_s)\frac{\partial u(t,t_s)}{\partial t} + k(t_s)\left(u(t,t_s) - \epsilon\, u^3(t,t_s)\right) = f(t,t_s)$$
$$(5.73)$$

This is a softening type nonlinear system and ϵ is a fixed constant which quantifies the "strength" of the nonlinearity. Here ϵ can also evolve with t_s. In that case, $\epsilon(t_s)$ needs to be identified from the sensor data. The establishment of this digital twin model therefore requires nonlinear dynamic analysis in the context of inverse problems.

6. *Digital twins for continuum systems*: Many engineering dynamic problems are modeled and subsequently analyzed using continuum models such as beams, plates and shells. The advantage of using a continuum model, as opposed to discrete or discretized models, is that it can give simple and physically insightful results. Digital twins for relevant real systems can make use of the continuum models for simplicity and practical relevance. As an example, digital twin of a damped Euler-Bernoulli beam [13] can be expressed as

$$EI(t_s)\frac{\partial^4 U(x,t,t_s)}{\partial x^4} + c_1(t_s)\frac{\partial^5 U(x,t,t_s)}{\partial x^4 \partial t} + m(t_s)\frac{\partial^2 U(x,t,t_s)}{\partial t^2}$$
$$+ c_2(t_s)\frac{\partial U(x,t,t_s)}{\partial t} = F(x,t,t_s) \quad (5.74)$$

In the above equation, x is the coordinate along the length of the beam, t is the time variable representing the beam dynamics, $EI(t_s)$ is the bending rigidity, $m(t_s)$ is the mass per unit, $c_1(t_s)$ is the strain-rate-dependent viscous damping coefficient, $c_2(t_s)$ is the velocity-dependent viscous damping coefficient, $F(x,t,t_s)$ is the applied spatial dynamic forcing and $U(x,t,t_s)$ is the transverse displacement. Similar to the SDOF system considered in this chapter, the digital twin will be realized by establishing the quantities which are functions of the slow time t_s. However, the solution of the equations is more complicated.

7. *Predictive response of digital twins*: Identifying the coefficient functions varying with t_s is the first step toward developing a digital twin for dynamic systems. A potential application of digital twin is for response predictions and subsequent engineering decisions. New efficient computational methods are needed for MDOF digital twins where the mass, damping and stiffness matrices can be large and also varying as functions of t_s. Moreover, if the forcing function is random, the dynamic response will be a random process evolving in t_s. This type of problem can be addressed using the framework of random vibration [139, 160]. New analytical formulations are needed to characterize random processes involving multiple time-scales.

8. *Uncertainty quantification for digital twins*: Quantification and management of uncertainties play a fundamental role in the construction of digital twins irrespective of its complexity, depth and range. Further research is necessary not only to model uncertainties in the coefficient functions and the forcing functions, but also to propagate such uncertainties in an efficient manner. The equation of motion of a stochastic MDOF digital twin is given by

$$\mathbf{M}(t_s,\xi)\frac{\partial^2 \mathbf{u}(t,t_s,\xi)}{\partial t^2} + \mathbf{C}(t_s,\xi)\frac{\partial \mathbf{u}(t,t_s,\xi)}{\partial t} + \mathbf{K}(t_s,\xi)\mathbf{u}(t,t_s,\xi) = \mathbf{f}(t,t_s,\xi)$$
$$(5.75)$$

where $\xi \in \Xi$ denotes the sample space. Therefore, $\mathbf{M}(t_s,\xi)$, $\mathbf{C}(t_s,\xi)$, and $\mathbf{K}(t_s,\xi)$ are $N \times N$ random matrices and $\mathbf{u}(t,t_s,\xi)$ and $\mathbf{f}(t,t_s,\xi)$ are

N dimensional random vectors. Stochastic finite element method can be adapted for use in systems with time-varying coefficients. Reduced order uncertainty propagation techniques such as polynomial chaos [82] and other surrogate modeling approaches [109, 147] can be applied for stochastic digital twins.

9. *Digital twins via Bayesian frameworks*: Bayesian approach typically start with a prior probability density function and then update it based on the available data. There are several theoretical and computational methods to achieve the Bayesian update. For example, several powerful filtering methods such as ensemble Kalman filter, extended Kalman filter and particle filter (see for example [69, 104]) have been proposed. Due to the availability of large data, often continuously in time t_s, digital twin technology is amenable toward Bayesian methods. There are significant prospects to apply new Bayesian methodologies to develop probabilistic models of digital twins. A Bayesian digital twin can generate more confidence in the users compared to their non-Bayesian counterparts.

10. *Machine learning and big data in digital twins*: Many modern systems come integrated with a large number of sensors. Moreover, these sensors often have a high sampling rate and data is continuously communicated from the system to the cloud where the information processing algorithms extract features from the data. This combination of a large number of sensors and a high sampling rate produces a plethora of data, leading to a big data problem. The data transmitted to the cloud needs to be used for updating the digital twin. The data is also contaminated with uncertainty from ubiquitous sources. Machine learning methods are useful in ensuring the digital twin matches the evolution of the physical system as closely as possible. Sensor data may be numerical, for example acceleration, strain or pressure measurements. However, data may also be in the form of text, speech and images obtained from recordings of noise from machinery, commands given by the pilot or maintenance engineer and images obtained by digital cameras/smart phones about changes in the system state. Deep learning methods can fuse big data and to extract features from it.

5.6 SUMMARY

The recent rise of digital twins motivated academic and industrial researchers to formalize and standardize underlying procedures and methodologies. However, a major drawback of such an approach is jargon-dominated literature, which many early adopters of digital twins may find it difficult to tailor for their specific application sector. Keeping this background in mind, this chapter proposed a specific but original approach to develop digital twins for structural dynamic systems. The physics-based model of a SDOF dynamic system is governed by second-order differential equations. The main scientific proposition is that the digital twin evolves at a time-scale which is much slower than the dynamics of the system. This makes it possible to identify critical system parameters as a function of the "slow-time" from continuously measured data. Based on the quantity, quality and the nature of the available data, two broad cases are considered. They include

1. Only the imaginary part of the complex natural frequency is available
2. Both the real and imaginary parts of the complex natural frequency are available

For each of these cases, three possible practical situations are envisaged, namely

1. Measured data is exact
2. Measured data contain explicit error
3. Measured data is available with error estimates

It is proposed in this chapter that the stiffness and mass of the SDOF digital twin evolve with the slow-time. Exact closed-form expressions are derived to establish the digital twin for these different cases.

The application of the new mathematical expressions has been illustrated using numerical examples. Simulated functions representing mass and stiffness variation as functions of the slow time have been developed considering aircraft systems. The situation when data is available at a lower sampling rate is investigated numerically. Although only SDOF dynamic systems are considered in this chapter, several conceptual extensions resulting from this work are described in detail. Some of these ideas will be explored more deeply in further chapters in this book. Others are left for readers of this book to explore.

6 Machine Learning and Surrogate Models

6.1 ANALYSIS OF VARIANCE DECOMPOSITION

Suppose $\mathbf{i} = (i_1, i_2, \ldots, i_N) \in \mathbb{N}_0^N$ be a multi-index with $|\mathbf{i}| = i_1 + i_2 + \cdots + i_N$. Now considering $\mathbf{x} = (x_1, x_2, \ldots, x_N)$ to be the random inputs, the unknown response $g(\mathbf{x})$ can be expressed as [178]:

$$g(\mathbf{x}) = \sum_{|\mathbf{i}|=0}^{N} g_{\mathbf{i}}(\mathbf{x_i})$$

$$= g_0 + \sum_{k=1}^{N} \sum_{i_1 < i_2 < \cdots < i_k} g_{i_1 i_2 \ldots i_k}(x_{i_1}, x_{i_2}, \ldots, x_{i_k})$$

(6.1)

where $g_{\mathbf{i}}(\mathbf{x_i})$ are the component functions.

Definition 1: The univariate terms (i.e., the terms corresponding to $k = 1$) in 6.1 are termed as first-order component functions. Similarly, the bivariate terms (terms corresponding to $k = 2$) are termed as second-order component functions. g_0 is the zeroth order component function.

Remark 1: First-order component function does not indicate linear variation and consists of terms having higher order cooperative effect. Order in ANOVA indicates number of cooperative terms involved in the expression.

Remark 2: The component functions defined in Equation 6.1 must be orthogonal to each other. This criteria is known as the *hierarchical orthogonality criteria*. Note that this is an essential condition to ensure uniqueness of the solution.

Now if, $x_1, x_2, \ldots, x_N$ are assumed to be independent, the component function over the problem space can be determined by imposing a vanishing condition [192]:

$$\int_{p_i}^{q_i} \varpi_k(x_k) g_{i_1 i_2 \ldots i_m}(x_{i_1}, x_{i_2}, \ldots, x_{i_m}) dx_k = 0, \quad \forall k \in \{i_1, i_2 \ldots, i_m\} \quad (6.2)$$

where ϖ_k denoted the PDF of x_k and p_i, q_i denotes the bounds of the variable. using Equation 6.2, the component function of ANOVA can be written as [113]:

$$g_0 = E(g(\mathbf{x}))$$
$$g_i(x_i) = E(g(\mathbf{x})|x_i) - g_0$$
$$g_{ij}(x_i, x_j) = E(g(\mathbf{x})|x_i, x_j) - g_i(x_i) - g_j(x_j) - g_0$$

(6.3)

$$\vdots$$

DOI: 10.1201/9781003268048-6

119

where $E(\bullet)$ denotes expectation. Classical ANOVA utilizes the equations specified in Equation 6.3 to determine the component functions. However, practical determination of the component function using classical ANOVA is computationally cumbersome and requires large number of training points. In order to address this issue, other variants of ANOVA have emerged. While the anchored ANOVA [47, 48] utilizes interpolation functions to represent the component function, the polynomial-based ANOVA [113, 118, 192] expresses the component functions in terms of some suitable basis. However, all the above mentioned ANOVA are only suitable for system involving only independent variables.

Of late, the concept of polynomial-based ANOVA has been extended for systems involving both correlated and independent random variables [93, 116]. This method, referred to as G-ANOVA express the component functions in term of extended bases. The unknown coefficients are determined by enforcing the orthogonality of the component functions. As already demonstrated, this method yields excellent result for high dimensional system [32, 33, 35, 41]. Here we discuss G-ANOVA as other version of ANOVA discussion is a subset of the same.

6.1.1 PROPOSED G-ANOVA

In this section, a new variant of G-ANOVA has been proposed. The basic idea is to represent the unknown component functions by using the polynomial chaos expansion (PCE). Utilizing the functional form of PCE, Equation 6.1 can be rewritten as:

$$g(\mathbf{x}) = g_0 + \sum_{|\mathbf{i}|=1}^{N} \sum_{|\mathbf{j_i}|=1}^{\infty} \alpha_{\mathbf{j_i}}^{\mathbf{i}} \psi_{\mathbf{j_i}} \tag{6.4}$$

Considering an Mth order ANOVA and r^{th} order PCE, Equation 6.4 reduces to:

$$\hat{g}(\mathbf{x}) = g_0 + \sum_{|\mathbf{i}|=1}^{M} \sum_{|\mathbf{j_i}|=1}^{r} \alpha_{\mathbf{j_i}}^{\mathbf{i}} \psi_{\mathbf{j_i}} \tag{6.5}$$

Equation 6.5 can be rewritten in matrix form as:

$$\Psi \alpha = \mathbf{d} \tag{6.6}$$

where Ψ is a matrix consisting of the orthogonal basis, vector α consists of the unknown coefficient vector and $\mathbf{d} = \mathbf{g} - \bar{\mathbf{g}}$ where, $\mathbf{g} = (g_1, g_2, \ldots, g_{N_S})^T$ is a vector consisting of the observed responses at N_S training points and $\bar{\mathbf{g}} = (g_0, g_0, \ldots, g_0)^T$ is the mean response vector. Premultiplying 6.6 by Ψ^T yields:

$$\mathbf{B}\alpha = \mathbf{C} \tag{6.7}$$

where $\mathbf{B} = \Psi^T \Psi$ and $\mathbf{C} = \Psi^T \mathbf{d}$. Close inspection of Ψ reveals identical columns. Thus, $\mathbf{B}$ has identical rows. These rows are redundants and can be removed. Removing identical rows of $\mathbf{B}$ and corresponding rows of $\mathbf{C}$, one obtains:

$$\mathbf{B}'\boldsymbol{\alpha} = \mathbf{C}' \tag{6.8}$$

Equation (6.8) represents a set of underdetermined equations and naturally there exists an infinite number of solutions. Assume $\mathbf{B}'$ to be a $p \times q$ matrix. Then all the solutions of Equation 6.8 can be represented as:

$$\boldsymbol{\alpha}(s) = (\mathbf{B}')^{-1}\mathbf{C}' + \left[\mathbf{I} - (\mathbf{B}')^{-1}\mathbf{B}'\right] v(s) \tag{6.9}$$

where $(\mathbf{B}')^{-1}$ denotes the generalized inverse of $\mathbf{B}'$, $v(s)$ is an arbitrary vector in $\mathbb{R}^q$ and $\mathbf{I}$ represents an identity matrix. One choice of $(\mathbf{B}')^{-1}$ in Equation 6.8 is $(\mathbf{B}')^{\dagger}$, where $\boldsymbol{\alpha}_0 = (\mathbf{B}')^{\dagger}\mathbf{C}'$ is the solution of obtained using the least-squared regression. Replacing $(\mathbf{B}')^{-1}$ by $(\mathbf{B}')^{\dagger}$ in Equation 6.9 yields:

$$\boldsymbol{\alpha}(s) = (\mathbf{B}')^{\dagger}\mathbf{C}' + \mathbf{P}v(s) \tag{6.10}$$

where

$$\mathbf{P} = \mathbf{I} - (\mathbf{B}')^{\dagger}\mathbf{B}' \tag{6.11}$$

Definition 2: Out of all the possible solution defined by 6.10, the solution that minimizes the least squared error and satisfies the hierarchical orthogonality criteria of G-ANOVA defined in Remark 2, is termed as the "best solution." In the proposed G-ANOVA, the "best solution," from all the available solution defined by Equation 6.10 is obtained by employing homotopy algorithm (HA) [34, 114, 119]. HA determines the unknown coefficient by minimizing the least-squared error and an objective function. The solution using HA is given as:

$$\boldsymbol{\alpha}_{HA} = \left[\mathbf{V}_{q-r}(\mathbf{U}_{q-r}^T\mathbf{V}_{q-r})^{-1}\mathbf{U}_{q-r}^T\right]\boldsymbol{\alpha}_0 \tag{6.12}$$

where $\mathbf{U}$ and $\mathbf{V}$ are matrices obtained by singular value decomposition of $\mathbf{PW}$ matrix:

$$\mathbf{PW} = \mathbf{U}\begin{pmatrix} \mathbf{A}_r & 0 \\ 0 & 0 \end{pmatrix}\mathbf{V}^T \tag{6.13}$$

For detailed derivation of 6.12, interested readers may refer [34, 114]. $\mathbf{P}$ in $\mathbf{PW}$ is the matrix defined in Equation 6.11 and $\mathbf{W}$ is the weight matrix used to formulate the objective function in HA. For details regarding the weight matrix $\mathbf{W}$, interested readers may refer [32, 116].

Once the unknown coefficient vector $\boldsymbol{\alpha}$ is determined, 6.4 provides an explicit mapping of the input and output variables. In this paper, the proposed G-ANOVA is utilized to generate an explicit expression for the limit-state function. Once determined, probability of failure P_f is calculated as [33]:

$$P_f = \frac{1}{2}\frac{\sqrt{\pi}\exp\left(\frac{1}{4}\frac{\lambda_1^2}{\lambda_2}\right)\left[\mathrm{erf}\left(\frac{1}{2}\frac{\lambda_1}{\sqrt{\lambda_2}}\right) - \mathrm{erf}\left(\frac{1}{2}\frac{2\lambda_2 y_l + \lambda_1}{\sqrt{\lambda_2}}\right)\right]\exp\left(-\lambda_0\right)}{\sqrt{\lambda_2}} \tag{6.14}$$

where erf $(\bullet)$ denotes error function. y_l in above equation represents the lower limit of response. The $\lambda_i's$, $i = 0, 1, 2$ in Equation 6.14 are functions of the first two statistical moments of the output response [33].

Remark 3: One advantage of the proposed G-ANOVA resides in the fact that it is possible to derive analytical formulas for the first two statistical moments. As a consequence, the moments obtained are free from the sampling error introduced when MCS is utilized for determining the statistical moments.

6.1.1.1 Statistical Moments

Lemma 1. The first moment of all but the zeroth order component function is zero.

Proof: An arbitrary mth order component function $g_{i_1 i_2 ... i_m} (x_{i_1}, x_{i_2}, ..., x_{i_m})$ can be represented as:

$$g_{i_1 i_2 ... i_m} (x_{i_1}, x_{i_2}, ..., x_{i_m}) = \sum_{|\mathbf{i}|=1}^{r} \alpha_{\mathbf{i}} \psi_{\mathbf{i}} (\mathbf{x_i}) \tag{6.15}$$

Applying the expectation operator $E (\bullet)$ on both sides:

$$E (g_{i_1 i_2 ... i_m}) = \sum_{|\mathbf{i}|=1}^{r} \alpha_{\mathbf{i}} E (\psi_{\mathbf{i}} (\mathbf{x_i})) \tag{6.16}$$

Now as already mentioned, $\psi_{\mathbf{i}}$ is a orthogonal polynomial and thus

$$E (\psi_{\mathbf{i}}) = 0, \quad \mathbf{i} = 1, 2, ..., r \tag{6.17}$$

Thus,

$$E (g_{i_1 i_2 ... i_m}) = 0 \tag{6.18}$$

From 6.18, it can be concluded that first moment of all but the zeroth order component function is zero.

Corollary 1. The mean of the proposed G-ANOVA is g_0.

Proof: 6.1 is rewritten as:

$$g (\mathbf{x}) = \underbrace{g_0}_{0^{\text{th}} \text{ order}} + \sum_i \underbrace{g_i (x_i)}_{1^{\text{st}} \text{ order}} + \sum_{1 \leqslant i < j \leqslant N} \underbrace{g_{ij} (x_i, x_j)}_{2^{\text{nd}} \text{ order}} + \cdots + \underbrace{g_{12...N} (x_1, x_2, ..., x_N)}_{N^{\text{th}} \text{ order}} \tag{6.19}$$

Applying expectation operator on both sides:

$$E (g (\mathbf{x})) = E (g_0) + \sum_i E (g_i (x_i)) + \sum_{1 \leqslant i < j \leqslant N} E (g_{ij} (x_i, x_j)) + \cdots$$
$$+ E (g_{12...N} (x_1, x_2, ..., x_N)) \tag{6.20}$$

Now g_0 is constant and hence $E (g_0) = g_0$. Furthermore from *Lemma 1*, expectation of all the other component function is zero. Thus,

$$E (g (\mathbf{x})) = g_0 \tag{6.21}$$

This completes the proof of *Corollary 1*.

Theorem 1. All the component functions are mutually uncorrelated.

Proof: Consider $\mathcal{J}$ and $\mathcal{K}$ to be two functions. From definition, $\mathcal{J}$ and $\mathcal{K}$ are orthogonal if

$$E(\mathcal{J}\mathcal{K}) = \frac{\mathbf{J}^T \mathbf{K}}{N_r} = 0 \tag{6.22}$$

where $\mathbf{J}$ and $\mathbf{K}$ are vectors consisting of N_r realization of the function $\mathcal{J}$ and $\mathcal{K}$. Similarly, $\mathcal{J}$ and $\mathcal{K}$ are uncorrelated if

$$E\left((\mathcal{J} - \bar{\mathcal{J}})(\mathcal{K} - \bar{\mathcal{K}})\right) = \frac{(\mathbf{J} - \bar{\mathbf{J}})^T (\mathbf{K} - \bar{\mathbf{K}})}{N_r} = 0 \tag{6.23}$$

From *Lemma 1*, the component functions (except the zeroth order) is having zero mean. Under this circumstances, Equations 6.22 and 6.23 are identical. Thus, one may conclude that all the component functions are mutually uncorrelated. This completes the proof of *Theorem 1*.

From *Theorem 1*, it is evident that covariance between any two component function is zero. Thus applying variance operator $\mathrm{var}(\bullet)$ on both sides of 6.19

$$\mathrm{var}(g(\mathbf{x})) = \mathrm{var}(g_0) + \sum_i \mathrm{var}(g_i(x_i)) + \sum_{1 \leqslant i < j \leqslant N} \mathrm{var}(g_{ij}(x_i, x_j)) + \cdots$$
$$+ \mathrm{var}(g_{12\ldots N}(x_1, x_2, \ldots, x_N)) \tag{6.24}$$

Now g_0 is constant and hence, $\mathrm{var}(g_0) = 0$. Thus, 6.24 reduces to

$$\mathrm{var}(g(\mathbf{x})) = \sum_i \mathrm{var}(g_i(x_i)) + \sum_{1 \leqslant i < j \leqslant N} \mathrm{var}(g_{ij}(x_i, x_j)) + \tag{6.25}$$
$$\cdots + \mathrm{var}(g_{12\ldots N}(x_1, x_2, \ldots, x_N))$$

Now considering an arbitrary mth order component function $g_{i_1 i_2 \ldots i_m}$ $(x_{i_1}, x_{i_2}, \ldots, x_{i_m})$ as before Equation (6.15) and applying the variance operator on both sides:

$$\mathrm{var}(g_{i_1 i_2 \ldots i_m}(x_{i_1}, x_{i_2}, \ldots, x_{i_m})) = \sum_{|\mathbf{i}|=1}^r (\alpha_{\mathbf{i}})^2 \, \mathrm{var}(\psi_{\mathbf{i}}(\mathbf{x_i})) \tag{6.26}$$

Using the property of orthogonal basis (6.17),

$$\mathrm{var}(\psi_{\mathbf{i}}(\mathbf{x_i})) = E\left((\psi_{\mathbf{i}}(\mathbf{x_i}))^2\right) - (E(\psi_{\mathbf{i}}(\mathbf{x_i})))^2$$
$$= E\left((\psi_{\mathbf{i}}(\mathbf{x_i}))^2\right) \tag{6.27}$$

Substituting 6.27 into 6.26,

$$\mathrm{var}(g_{i_1 i_2 \ldots i_m}(x_{i_1}, x_{i_2}, \ldots, x_{i_m})) = \sum_{|\mathbf{i}|=1}^r (\alpha_{\mathbf{i}})^2 E\left((\psi_{\mathbf{i}}(\mathbf{x_i}))^2\right) \tag{6.28}$$

Furthermore if special category of orthogonal polynomial, known as the orthonormal polynomial is used,

$$E\left(\left(\psi_{\mathbf{i}}\left(\mathbf{x_i}\right)\right)^2\right) = 1 \tag{6.29}$$

Thus,

$$\text{var}\left(g_{i_1 i_2 \ldots i_m}\left(x_{i_1}, x_{i_2}, \ldots, x_{i_m}\right)\right) = \sum_{|\mathbf{i}|=1}^{r}\left(\alpha_{\mathbf{i}}\right)^2 \tag{6.30}$$

Substituting Equation 6.30 into 6.25 and writing using the multi-index notation

$$\text{var}\left(g\left(\mathbf{x}\right)\right) = \sum_{|\mathbf{i}|=1}^{N}\sum_{|\mathbf{j_i}|=1}^{\infty}\left(\alpha_{\mathbf{j_i}}^{\mathbf{i}}\right)^2 \tag{6.31}$$

and thus,

$$E\left(\left(g\left(\mathbf{x}\right)\right)^2\right) = \left(g_0\right)^2 + \sum_{|\mathbf{i}|=1}^{N}\sum_{|\mathbf{j_i}|=1}^{\infty}\left(\alpha_{\mathbf{j_i}}^{\mathbf{i}}\right)^2 \tag{6.32}$$

which (6.32) is the basic formula for calculating the second moment of the proposed G-ANOVA. Once the first two moments are determined using Equations 6.31 and 6.32 respectively, the procedure described in [33] is employed to determine the λ's. Finally, Equation 6.14 is employed to determine the probability of failure.

6.2 POLYNOMIAL CHAOS EXPANSION

The polynomial chaos expansion is another effective meta-modeling approach that is commonly utilized for uncertainty quantification (PCE). Wiener [206] was the first to develop this metamodel, hence it's also known as the "Wiener Chaos expansion". Further, Xui [215] went on to generalize the findings from the so-called Askey-scheme for numerous continuous and discrete systems. The work also presents the occurrence of convergence in the associated Hilbert space. The fundamental concept behind reliability analysis utilizing PCE will be described below.

If it is assumed that $\mathbf{i} = (i_1, i_2, i_3, \ldots, i_n) \in \mathbb{N}_0^n$ to be a multi-index with $|i| = i_1 + i_2 + i_3 + \ldots + i_n$ and $N \geq 0$, Then N^{th} order PCE is given by

$$g(\hat{\mathbf{Z}}) = \sum_{|i|=0}^{N} a_i \Phi_i(\mathbf{Z}) \tag{6.33}$$

where, $\{a_i\}$ are the unknown coefficients and $\Phi_i(Z)$s are the n-dimensional polynomials satisfies the orthogonality conditions given by

$$\langle \Phi_i(Z)\Phi_j(Z)\rangle = \int_{\Omega}\Phi_i(Z)\Phi_j(Z)dF_z(Z) = \delta_{ij} \tag{6.34}$$

Table 6.1

The Correspondence of the Type of Orthogonal Polynomial with Distribution Pattern

Distribution	Random variable	Polynomial	Support
Continuous	Gaussian	Hermite	$(-\infty, \infty)$
	Gamma	Lagurre	$[0, \infty)$
	Beta	Jacobi	$[a, b]$
	Uniform	Legendre	$[a, b]$
Discrete	Poisson	Charlier	$\{0, 1, ...\}$
	Binomial	Krawtchouk	$\{0, 1, ..., N\}$
	Negative Binomial	Meixner	$\{0, 1, ...\}$
	Hypergeometric	Hahn	$\{0, 1, ..., N\}$

where, $0 \leq |i|, |j| \leq N$ and δ_{ij} denotes the multivariate kronecker function. It should be noted that according to the probability space of the random variable (dF_z), the corresponding orthogonal polynomials are chosen (see Table 6.1). The correspondence orthogonal polynomial with distributions patterns are given in the table. There are different approaches proposed in the literature to determine the unknown coefficients. Out of the methods popularly employed methods are least square method as well as collocation method. Once the model is approximated, similar to the other surrogate methods, Monte Carlo simulation (MCS) is performed to determine the failure probability. Furthermore, because of the *curse of dimensionality*, this approach is unfeasible for high stochastic dimensions.

6.3 SUPPORT VECTOR MACHINES

Support Vector Machine (SVR) model is a unique learning method that was originally employed in pattern recognition tasks. However, later on, it has been utilized for the regression-based tasks as it performs equally well compared to the other existing methods. Training approach entails obtaining the correlation or nonlinear mapping function $f(x)$ between the input and output, and the output of the learner. Also, SVR seeks to give a nonlinear mapping function to map the training data $\{x_i, y_i\}$ with $i = 1, ..., n$ to a high-dimensional feature space. Then, it is possible to characterize the relationship between the input and output of the learner by using the following regression function:

$$f(\boldsymbol{X}) = \boldsymbol{W}^T \phi(\boldsymbol{X}) + b \tag{6.35}$$

where W and b are unknown coefficients that are to be determined. Further, empirical risk can be expressed as follows:

$$R_{emp}(f) = \frac{1}{n} \sum_{i=1}^{n} \Theta_\varepsilon(y_i, f(X)) \tag{6.36}$$

Θ_ε in the above expression is ε-intensive loss function and can be defined as follows:

$$\Theta_\varepsilon(y_i, f(\boldsymbol{X})) = \begin{cases} |f(X)-y|-\varepsilon, & \text{if } \Theta_\varepsilon(y_i, f(\boldsymbol{X})) \geq \varepsilon \\ 0, & \text{otherwise} \end{cases} \tag{6.37}$$

SVR obtains the hyper planes which divides the training data into linearly separable subsets with maximum separation distance through optimization where the objective function is given by:

$$Min_{w,b,\xi^*,\xi} R_\varepsilon(w, \xi^*, \xi) = \frac{1}{2} w^T w + C \sum_{i=1}^{n} (\xi_i + \xi_i^*) \tag{6.38}$$

where C is the trade-off parameter between the equation's first and second terms. In order to regularize the huge weights in this equation, they are modified by maximizing the distance. Moreover, between two different sets of data, the ε-insensitive loss function is used to penalize the f(x) and y training error. The following are the constraints of this optimization problem:

$$\begin{aligned} y_i - w^T \varphi(X_i) - b &\leq \xi_i^* \varepsilon, i = 1, 2, 3, ..., n \\ -y_i + w^T \varphi(X_i) + b &\leq \xi_i^* \varepsilon, i = 1, 2, 3, ..., n \\ \xi_i^* \varepsilon &\geq 0, i = 1, 2, 3, ..., n \end{aligned} \tag{6.39}$$

The coefficients w can be determined by solving the optimization problem and can be expressed as follows:

$$w = \sum_{i=1}^{n} (\beta_i - \beta_i^*) \varphi(X_i) \tag{6.40}$$

Here β_is are the Lagrangian coefficients. Finally, the SVR regression function is described as:

$$f(X) = \sum_{i=1}^{n} (\beta_i - \beta_i^*) K(X_i, X_j) + b \tag{6.41}$$

Here, $K(X_i, X_j)$ represents the Kernel function. In feature space Kernel function is defined as:

$$K(X_i, X_j) = \varphi(X_i).\varphi(X_j) \tag{6.42}$$

The most commonly used Kernel functions are Gaussian Radial Basis Function (RBF) and polynomial.

6.4 NEURAL NETWORKS

An Artificial Neural Network (ANN) is an information processing paradigm that is inspired from biological nervous systems. The frameworks are now used for a wide range of computational tasks. ANN and *deep learning* have been recognized as a potential solution in a broad range of applications including computer vision, natural language processing, Internet of Things (IoT), speech processing, neuroscience, autonomous driving car, and so on.

Neocognitron[72] in 1980 considered as the first mathematical model of neural network, which possess some of the characteristics of the convolutions network. According to the universal approximation theorem[54], any arbitrary function can be represented as neural network with adequate number of hidden layers and neurons. A schematic representation of an ANN is shown in Figure 6.1. Deep Convolutional Neural Network(DCNN) [110] are one of the potential category of NNs, have revolutionized the field of computer vision by dominating the performance metrics in almost every meaningful computer vision task. NN has powerful representative capability and thus the auto encoder algorithm [46] and its deep counterpart have gained significant success as traditional dimensionality reduction approaches. ANN can be used as an surrogate model to approximate the complex implicit performance functions. However, the applications of the ANN is not limited to only surrogate model. The basic entity of any NN is known as neuron and the neurons need to be connected with each other and forms the neural network. Feed-forward structure is one of the simplest and widely used architecture.

The basic idea of a neuron model is that for a given data set $\{x_i, y_i\}_{i=1}^{N}$ an input, x_i, together with a bias, b is weighted by, w, and then summarized together to get the input to the next neuron $a_m = \sum w_{lm} x_l + b_l$, known as activation value. Output to the neuron is a linear or nonlinear functional map of this activation value $Z_m = f(a_m)$. Similarly, the output of the next neurons is passed to the successive neurons. Finally, the approximation of the y can be obtained as:

$$\hat{y} = f_\theta(\mathbf{m}x_i) \tag{6.43}$$

Network is trained using the error function $\ell(\hat{y}, y_i)$, where one has to appropriately choose the error function. During training, error functions are reduced by optimizing the parameters θ_i

$$\boldsymbol{\theta}^* = \arg\min_{\boldsymbol{\theta}} \sum_{i=1}^{N} \ell\left(f_{\boldsymbol{\theta}}\left(x_i\right), y_i\right), \tag{6.44}$$

where $\theta_i = \{w_i, b_i\}$ A schematic diagram of feed forward NN is shown in the Figure 6.1. NN outperforms in learning the complex nonlinear relationship of input to output data. It has proven to be a powerful tool in the domain of image processing. However, ANN do not guarantee that the actual physics of the problem is satisfied, while dealing with systems which are driven by certain governing lows. Moreover, it fails to give accurate results when the enough training data is not available. To overcome the shortcoming of the ANN researchers have developed physics constrained NN, also known as *Physics Informed Neural Network*, (PINN). Numerous studies on PINN are found in the literature [80, 102, 149, 150] over the past few years. The basic idea here is to train the NN constraining the governing physics, when the given law of physics described as general nonlinear differential equations.

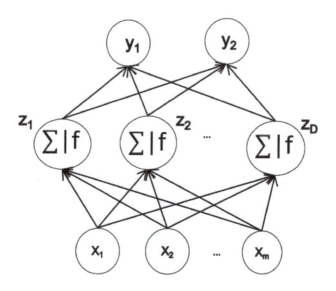

Figure 6.1 Schematic of neural network

6.5 GAUSSIAN PROCESS

Gaussian process models have gained wide attention for solving non linear regression as well as classification problems. Gaussian Process Regression (GPR) avoids over-fitting, as most of the available regression methods are prone to over-fitting and it also has impressive predictive performance in many thorough empirical comparisons. GP models allow exact Bayesian analysis with simple matrix manipulations and provide good performance. The basic concept behind the GPR is described below.

Consider a finite set of input data $X = x_1, x_2, x_3, ...x_n$, having a covariance matrix $K_{ij} = k(x_i, x_j)$. $Y = y_1, y_2, y_3, ...y_n$, we assume that for each input x there is a corresponding output $y(x)$. The input to output relation can be expressed as follows:

$$y = t(x) + \xi \tag{6.45}$$

where $t(x)$ is the random variable with $t = (t(x_1), t(x_2), ...t(x_n))^T \sim \mathcal{N}(0, K)$ and ξ is random variable with $\xi \sim \mathcal{N}(0, \sigma^2)$. Further, distribution of the output $y(x)$ can be determined by Bayes theorem. Conditional distribution of $y(x)$ on given input output data is normal. Equivalent parametric representation of $y = (y(x_1), y(x_2), ..., y(x_n))^T$ is expressed as:

$$y = \mathbf{K}\alpha + \xi \tag{6.46}$$

where $\alpha \sim \mathcal{N}(0, K^{-1})$, and $\xi \sim \mathcal{N}(0, \sigma^2 \mathbf{I})$. Thus posterior probability $p(\alpha/\boldsymbol{y}, \mathbf{X})$

$$p(\alpha/\boldsymbol{y}, \mathbf{X}) \propto \exp(-\frac{1}{2\sigma^2}|y - K\alpha| + \frac{1}{2}) \exp(-\frac{1}{2}\alpha^T K\alpha) \qquad (6.47)$$

Conditional expectation of new point x is given by

$$E[y(x)/y, \mathbf{X}] = \mathbf{k}^T \alpha_{opt} \qquad (6.48)$$

where $\mathbf{k}^T$ denotes $\mathbf{k}^T = (k(x_1, x), k(x_2, x), ..., k(x_n, x))$ and α_{opt} is the value of α that maximises the $p(\alpha/\boldsymbol{y}, \mathbf{X})$. Moreover, maximum a posteriori (MAP) estimate of α can be calculated by minimizing the negative log-posterior, and which is given by

$$\alpha_{MAP} = \text{minimize}[-y^T K\alpha + \frac{1}{2}\alpha^T(\sigma^2 K + K^T K)\alpha], \quad \alpha \in R^M \qquad (6.49)$$

Now mean and variance of the cognitional distribution is give by $k^T(K + \sigma^2 I)^{-1}y, k(x, x) + \sigma^2 - k^T(K + \sigma^2 I)^{-1}k$ respectively.

6.6 HYBRID POLYNOMIAL CORRELATED FUNCTION EXPANSION

Hybrid polynomial correlated function expansion (H-PCFE) can be viewed as an improvement over the conventional GP where the mean function of the GP is represented by using the functional form of PCFE.

$$\mu_f(\boldsymbol{\xi}; \boldsymbol{\beta}) \approx g_0 + \sum_{k=1}^{M_o} (\sum_{i_1=1}^{N-k+1} \sum_{i_2=i_1}^{N-k+2}$$
$$\cdots \sum_{i_k=i_{k-1}}^{N} \sum_{r=1}^{k} \left(\sum_{m_1=1}^{s} \cdots \sum_{m_r=1}^{s} \beta_{m_1...m_r}^{(i_1 i_2...i_k)i_r} \Psi_{m_1}^{i_1} \cdots \psi_{m_r}^{i_r} \right)).$$

where ψ are the orthonormal basis function and β represents the unknown coefficients associated with the bases. g_0 in Equation 6.50 is a constant and termed as the zeroth order component function. In Equation 6.50, we have used a M-th order PCFE with s-th order basis function.

For computing the hyperparameters of the H-PCFE, we utilize the maximum likelihood estimator (MLE)

$$\left(\boldsymbol{\Psi}^T \mathbf{R}^{-1} \boldsymbol{\Psi} \right) \boldsymbol{\beta} = \boldsymbol{\Psi}^T \mathbf{R}^{-1} \boldsymbol{J}_s, \qquad (6.50a)$$

$$\sigma^2 = \frac{1}{N_s} \left(\boldsymbol{J}_s - \boldsymbol{\Psi}\boldsymbol{\beta} \right)^T \mathbf{R}^{-1} \left(\boldsymbol{J} - s - \boldsymbol{\Psi}\boldsymbol{\beta} \right), \qquad (6.50b)$$

where $\boldsymbol{\Psi} \in \mathbb{R}^{N_s \times N_b}$ and $\mathbf{R} \in \mathbb{R}^{N_s \times N_s}$ are respectively the basis function matrix (i.e., design matrix) and the covariance matrix is computed based on the training samples. $\boldsymbol{J} \in \mathbb{R}^{N_s}$ is the response vector and $\boldsymbol{J}_s = \boldsymbol{J} - g_0$ is the *shifted* response vector corresponding to the training samples. Note that there exists no closed form solution for the length scale parameter $\boldsymbol{\theta}$ and hence, we have no option but to solve a numerical optimization problem. In this work, stochastic gradient descent has been used for computing $\boldsymbol{\theta}$. We have assumed that the training is carried out corresponding to N_s training samples.

We note that computing $\boldsymbol{\beta}$ from Equation 6.50a is non trivial; this is because $\mathbf{A} = \boldsymbol{\Psi}^T \mathbf{R}^{-1} \boldsymbol{\Psi}$ has identical rows. Removing the identical rows results in an under-determined matrix $\mathbf{A}'$. In order to solve this under-determined set of equations, we have utilized a homotopy algorithm (HA) [115]. The advantage of HA resides in the fact that it is possible to satisfy the hierarchical orthogonality criterion associated with PCFE and H-PCFE [117]. The optimal $\boldsymbol{\beta}$ obtained using HA are as follows:

$$\beta_{HA} = \left[\mathbf{V}_{N_b-k} \left(\mathbf{U}_{N_b-k}^T \mathbf{V}_{N_b-k} \right)^{-1} \mathbf{U}_{N_b-k}^T \right] \beta_0, \qquad (6.51)$$

where

$$\beta_0 = (\mathbf{A}')^\dagger \, \mathbf{D}'. \qquad (6.52)$$

$\mathbf{D}'$ and $\mathbf{A}'$ in Equation 6.52 are respectively $\mathbf{D} = \boldsymbol{\Psi}^T \mathbf{R}^{-1} \mathbf{J}_s$ and $\mathbf{A}$ after removing the identical rows. $(\mathbf{A}')^\dagger$ represents the pseudo inverse of $\mathbf{A}'$ that satisfies also four Penrose conditions [152]. $\mathbf{U}$ and $\mathbf{V}$ in Equation 6.51 are obtained as

$$\mathbf{PW}_{HA} = \mathbf{U} \begin{bmatrix} \mathbf{S}_k & \mathbf{0} \\ \mathbf{0} & \mathbf{0} \end{bmatrix} \mathbf{V}^T, \qquad (6.53)$$

where $\mathbf{P} = \left(\mathbb{I}_{N_b} - (\mathbf{A}')^\dagger \mathbf{A}' \right)$ and $\mathbf{W}_{HA}$ is a weight matrix that is used in HA to ensure the hierarchical orthogonality criterion. Details on the weight matrix used in this study can be found in [38]. $\mathbf{U}_{N_b-k}$ and $\mathbf{V}_{N_b-k}$, respectively, denotes the last $N_o - k$ columns of $\mathbf{U}$ and $\mathbf{V}$.

$$\mathbf{U} = [\mathbf{U}_k \mathbf{U}_{N_b-k}], \qquad \mathbf{V} = [\mathbf{V}_k \mathbf{V}_{N_b-k}]. \qquad (6.54)$$

For further details of the HA, interested readers may refer [115]. For further details on the training phase of H-PCFE, interested readers may refer [36]. An algorithm depicting the steps involved in training an H-PCFE model is shown in Algorithm 2.

Once the training phase is complete and we have obtained the hyperparameters of the H-PCFE model, we proceed to the prediction phase. In this phase, the goal is to predict the responses J_{pred} corresponding to unobserved input $\boldsymbol{\xi}_{pred}$. To that end, we compute the predictive distribution of response

$$\mathbb{P}\left(J_{pred} \,|\, \boldsymbol{\xi}_{pred}, \boldsymbol{\xi}, \boldsymbol{J}, g_0\right) = \mathcal{N}\left(J_{pred} \,|\, \mu_{pred}, \sigma_{pred}\right), \qquad (6.55)$$

where

$$\mu_{pred} = g_0 + \boldsymbol{\Psi}_{pred}\beta_{HA} + r\mathbf{R}^{-1}\left(\boldsymbol{J}' - \boldsymbol{\Psi}_{pred}\beta_{HA}\right), \qquad (6.56)$$

Algorithm 2 H-PCFE: training

Pre-requisite: Training samples $\boldsymbol{\xi}^{(i)}$, $\boldsymbol{J}^{(i)}$, $i = 1, \ldots, N_s$. Provide the order of PCFE/H-PCFE m, maximum degree of basis function s and form of the correlation function R.

Obtain the length scale parameters, $\boldsymbol{\theta}$ by maximizing the likelihood.

Formulate the design matrix $\boldsymbol{\Psi}$.

Formulate the correlation matrix $\mathbf{R}$.

Formulate the weight matrix $\mathbf{R}$.

Formulate weight matrix, $\mathbf{W}_{HA}$ to be used in HA [38].

Compute $\mathbf{U}$ and $\mathbf{V}$ using 6.53.

Compute β_0 using 6.52.

Compute β_{HA} using 6.51. Compute the process variance σ^2 using 6.50b.

Outcome: H-PCFE hyperparameters, β_{HA}, $\boldsymbol{\theta}$ and σ^2.

---

and

$$\sigma_{pred} = \sigma^2 \left\{ 1 - \boldsymbol{r}^T + \frac{1 - \boldsymbol{\Psi}^T \mathbf{R}^{-1} \boldsymbol{r}}{\boldsymbol{\Psi}^T \mathbf{R}^{-1} \boldsymbol{\Psi}} \right\}. \tag{6.57}$$

$\boldsymbol{\Psi}_{pred}$ and $\boldsymbol{r}$ in Equations (6.56) and (6.57) are, respectively, the basis function vector evaluated at $\boldsymbol{\xi}_{pred}$ and a vector representing the correlation between the prediction point $\boldsymbol{\xi}_{pred}$ and the training points.

7 Surrogate-Based Digital Twin of Dynamic System

As we discussed in Chapter 1, a digital twin is a virtual model of a physical system which exists in the computer cloud. Increasingly, this physical system is called a physical twin [180]. Attempts to simulate the behavior of physical systems are a crucial component of engineering and science. The difference between the digital twin and a computer model is that the digital twin updates itself to track the physical twin through the use of sensors, data analysis, signal processing, and the internet of things (IoT). The digital twin may also direct changes to the physical twin through signals sent to actuators on the physical twin. An ideal digital twin possesses temporal synchronization with respect to the physical twin.

The concept of the digital twin is broad; it can be used for mechanical and aerospace systems, manufacturing processes, smart cities and biological systems. A digital twin is born as a nominal model of a system when it begins its operational life. At $t = t_s = 0$, the digital and the physical twins are identical. Thereafter, the digital twin of this system tracks its evolution over time. As discussed in Chapter 1, the real system and its digital twin are subject to at least two-time scales. The first time scale is of instantaneous time t, and the second time scale is of "slow time" t_s which tracks the evolution of the properties of the real system. For example, in gas turbine diagnostics, the module efficiencies degrade slowly as the engine is placed in service with a linear change with respect to t_s being a good approximation [73]. Damage growth in composites occurs over time due to fatigue and can be represented as a curve fit [125]. This curve fit with respect to t_s captures the physical damage modes of matrix cracking, delamination and fibre breakage.

The ubiquitous connectivity of physical systems to the internet permits the tracking of millions of physical twins in real-time. However,it is challenging to ensure that the digital twin mimics the physical system closely with respect to time. In other words, the digital twin must evolve with the physical twin over its life cycle. Thus, the digital twin must be subject to the same conditions or nurture as the phsyical twin in terms of environment, loads, maintainance, repair, damage, etc. At the end of life, both digital and physical twins may be terminated. On the other hand, the digital twin may be allowed to continue as a repository of information about the expired physical twin. The choice will depend on the rquirements of the system.

The physical twin exists in the real world and the digital twin exists in the cloud [52]. To recapitulate, problems in the development of digital twin technology are related to

DOI: 10.1201/9781003268048-7

1. Modeling and simulation of the physical twin
2. Accuracy and speed of data transmission between the physical twin and the cloud where the digital twin resides
3. Storage and processing of big data created by sensors through the life cycle
4. Efficient computer processing of the digital twin model in the cloud
5. Plausible actuation of the physical twin through signals sent by the digital twin from the cloud to actuators on the physical twin

An early paper on digital twins in the context of life prediction of aircraft structures was published by Tuegel and his co-workers [187]. Their objective was to apply an ultra-high fidelity model of an individual aircraft identified by its tail number to predict its remaining structural life. This research was motivated by the evolution of high-performance computing. The key idea introduced in this paper was the need to track *each individual aircraft by its tail number, which is a unique identifier*. Thus, every physical aircraft in the fleet would have a separate digital twin, which will evolve differently compared to the other twins as each aircraft faces unique combinations of missions, loads, environment, pilot behavior, sensor situation, etc. This concept of digital twins was later expanded to any physical system such as cars, computer servers, locomotives, turbines, machine tools, etc. The primary applications of digital twin have occured in production, product design, and prognostics and health management [88, 89, 124, 182, 226]. However, a plethora of applications are possible in the future. Modern tagging methods such as bar codes and IP addresses for "things," national identification numbers for humans, image and face recognition technology, and machine learning enable us to create digital twins of the man-machine ecosystems using ubiquitous wi-fi available in smart cities and smart phone-based GPS.

Digital twin technology has found several industrial applications, as adumbrated by Tao et al. in their recent review paper [183]. According to Tao and his co-researchers, digital twin is an enabling technology for facilitating smart manufacturing and Industry 4.0. In Industry 4.0, factories are envisaged with wireless connectivity and sensors which permit the production line to function automatically. Since physical systems are increasingly embellished with sensors, data transmission technology permits the collection of data throughout the life stage of the physical system. This data set is enormous, and big data analytics is required to discover failure causes, streamline supply chains and enhance production efficiency. Prognostics and health management is the area which has witnessed considerable application of digital twin technology. Li et al. [218] constructed a digital twin based on the dynamic Bayesian network to monitor the operational state of aircraft wings. A probabilistic digital twin model was used to emulate the deterministic physical model. Digital twins have also been applied in prognostics and health management of cyber-physical systems and additive manufacturing processes. The fusion of the physical and the cybernetic systems is the key challenge for the use of digital twins, and one approach to address this problem is to make the modeling and simulation tasks more efficient. Surrogate modeling used in this chapter could be one approach to increase the efficiency of the simulation process.

Singh and Wilcox [176] discuss the related concept of digital thread in the context of engineering design. They mention that a digital twin can be considered as a "high fidelity digital representation to closely mirror the life of a particular product and serial number (e.g., loading history, part replacement, damage etc." The digital thread contains all the information needed to generate and provide updates to a digital twin. They adumbrate an example to clarify the digital thread concept. Consider a mechanical component in service which is equipped with strain sensors to identify the loading conditions. The next generation of the product design could utilize this strain data to better model the loading condition in operation and therefore improve the future design of the component. Therefore, information and resources from different stages of the product lifecycle must be fed back to the design process. This process is completed by the digital thread. For the problem selected, the digital thread consists of the probabilistic distributions of the input variables, the finite element solver and failure criteria used for modeling, the noise statistics of the measurements and the geometry of the components of the system. The probabilistic distributions of the uncertain input variables are estimated using the Kalman filter. Uncertainty quantification is important for the development of digital twin and digital thread technology.

In Chapter 5, a physics-based digital twin model for a dynamical system was developed. The theory was based on the assumption that the properties of the physical system evolve with much lesser alacrity than real-time. This system property evolution takes place in slow time rather than real-time. A single-degree-of-freedom (SDOF) system model was created to explain the concept of digital twin and study the effect of changes in system mass and stiffness properties. Closed-form solutions for the evolution of mass and stiffness with the slow time were proposed. For clean data, the proposed closed-form solutions yielded exact estimates of the mass and stiffness evolution. However, the proposed physics-based digital twin has two major limitations.

1. The physics-based digital twin only yields the mass and stiffness at the slow time-step; i.e., the time-steps at which sensor measurements exist. Therefore, the physics-based digital twin is incapable of providing the mass and stiffness at the intermediate and future time.
2. The physics-based digital twin can fail if the data collected is noise contaminated. This is a major limitation since in real-life, collected data is always corrupted by noise.

Thus, there remains a lack of clarity and specificity regarding physics-based digital twins. In this chapter, the concept of surrogate models within the digital twin framework is introduced. This is motivated by the fact that a physics-driven digital twin, such as the one proposed in Chapter 5 only provides estimates at discrete time-steps (the time-steps corresponding to sensor measurements). Moreover, such physical model-based digital twins may yield erroneous results when the data collected is corrupted by noise. By definition, a surrogate model is a *proxy* to the physics based high-fidelity model. Popular surrogate models available in the literature include analysis-of-variance decomposition [39], polynomial chaos expansion [181, 214], support vector machines [22, 87, 224], neural networks [27, 28, 85], and Gaussian process (GP) [18, 19, 40]. Successful use of surrogate models is observed in several domains including, but not limited to, stochastic mechanics [60, 211], reliability analysis [37, 65] and optimization [86, 101, 148]. This chapter illustrates how surrogate models can be integrated with the physics-based digital twin technology and the advantages of this integration.

All the surrogate models discussed above can be used within the digital twin framework. This chapter explores the use of GP, a probabilistic machine learning technique that seeks to infer a distribution over functions and then makes predictions at some unknown points. There are two main advantages of GP over other surrogate models.

1. GP is a probabilistic surrogate model and is immune to over-fitting. This is crucial for the digital twin as the data collected are corrupted by complex noise and over-fitting can result in erroneous predictions.
2. GP can quantify the uncertainty emanating from limited and noisy data. This, in turn, can be applied in the decision-making process using a digital twin.

In this chapter, we have used *vanilla* GP as a surrogate to develop the data-driven digital twin model.

The chapter is organized as follows. In Section 7.1, the equation of motion of an SDOF digital twin is developed using multiple time-scales. The fundamentals of GP are discussed in Section 7.2. The development of a digital twin using only the mass evolution is considered in Section 7.3. Three separate cases with (a) only mass evolution, (b) only stiffness evolution, and (c) both mass and stiffness evolution are considered. Numerical examples are given to showcase the proposed ideas. A critical discussion on the proposed methodology as well as the overall development of digital twin is given in Section 7.4 and concluding remarks are proferred in Section 7.5. This chapter is adapted from Ref. [30].

7.1 THE DYNAMIC MODEL OF THE DIGITAL TWIN

A SDOF dynamic system is again considered to explore the concept of a digital twin. A key idea is that a digital twin starts its journey from a "nominal model." The nominal model is, therefore, the "initial model" or the "starting model" of a digital twin. For dynamic systems, the nominal model is often a physics-based model which has been verified, validated, and calibrated. For example, the nominal model can be a finite element model of a car, bridge, ship, turbine, or an aircraft when the product leaves the manufacturer facility and is ready to go into service. Another key characteristic of a digital twin is its connectivity. Internet of Things (IoT) brings forward numerous new data technologies and accelerates the development of digital twin technology. IoT enables connectivity between the physical SDOF system and its digital counterpart. The basis of digital twins is based on this connection; without it, digital twin technology would be hard to fructify. The connectivity is facilitated by sensors on the physical system which obtain data and integrate and communicate this data through various technologies.

Consider a physical system which can be well approximated by a single degree of freedom spring, mass, and damper system, as before [77]. Assume that the sensors sample data intermittently, typically, t_s represents discrete time points. Assume that changes in $k(t_s)$, $m(t_s)$, and $c(t_s)$ as so slow that the dynamics of the system is effectively decoupled from these functional variations. The equation of motion is then

$$m(t_s)\frac{\partial^2 u(t,t_s)}{\partial t^2} + c(t_s)\frac{\partial u(t,t_s)}{\partial t} + k(t_s)u(t,t_s) = f(t,t_s) \qquad (7.1)$$

Here t and t_s are the system time and a "slow time," respectively. $u(t,t_s)$ is a function of two variables and therefore the equation of motion is expressed in terms of the partial derivative with respect to the time variable t. The slow time or the service time t_s can be considered as a time variable which is much slower than t. For example, it could represent the number of cycles in an aircraft. Thus, mass $m(t_s)$, damping $c(t_s)$, stiffness $k(t_s)$ and forcing $F(t,t_s)$ change with t_s, for example due to system degradation during its service life. The forcing is also a function of time t and slow time t_s, as is the system response $x(t,t_s)$. Equation (7.1) is considered as a digital twin of a SDOF dynamic system. When $t_s = 0$, that is at the beginning of the service life of the system, the digital twin Equation (7.1) condenses to the nominal system

$$m_0\frac{\mathrm{d}^2 u_0(t)}{\mathrm{d}t} + c_0\frac{\mathrm{d}u_0(t)}{\mathrm{d}t} + k_0 u_0(t) = f_0(t), \qquad (7.2)$$

where m_0, c_0, k_0, and f_0, are respectively the mass, damping, stiffness and force at $t = 0$. It is assumed that sensors are deployed on the physical system and take measurements at locations of time defined by t_s. The functional form of the relationship of mass, stiffness and forcing with t_s is unknown and should be estimated from measured sensor data. We assume that damping is small so that the effect of variations in $c(t_s)$ are negligible. Thus, only variations in the mass and stiffness are considered. Without any loss of generality, the following functional forms are considered

$$k(t_s) = k_0(1 + \Delta_k(t_s))$$
$$\text{and} \quad m(t_s) = m_0(1 + \Delta_m(t_s)) \qquad (7.3)$$

Typically, $k(t_s)$ is expected to be a decaying function over a long time to represent a loss in the stiffness of the system. On the other hand, $m(t_s)$ can

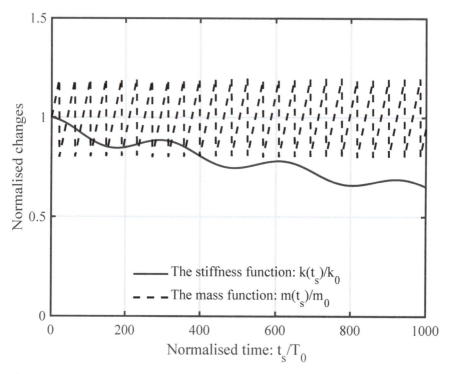

Figure 7.1 Examples of model functions representing long term variabilities in the mass and stiffness properties of a digital twin system.

be an increasing or a decreasing function. For example, for an aircraft, mass can represent the loading of cargo and passengers and also represent the use of fuel as the flight progresses. The following representative functions have been chosen as examples

$$\Delta_k(t_s) = e^{-\alpha_k t_s} \frac{(1 + \epsilon_k \cos(\beta_k t_s))}{(1 + \epsilon_k)} - 1 \qquad (7.4)$$

$$\text{and} \quad \Delta_m(t_s) = \epsilon_m \, \text{SawTooth}(\beta_m(t_s - \pi/\beta_m)) \qquad (7.5)$$

Here SawTooth($\bullet$) represents a sawtooth wave with a period 2π. In Figure 7.1 overall variations in stiffness and mass properties arising from these function models have been plotted as function of time normalized to the natural time period of the nominal model. Numerical values used for these examples are $\alpha_k = 4 \times 10^{-4}$, $\epsilon_k = 0.05$, $\beta_k = 2 \times 10^{-2}$, $\beta_m = 0.15$ and $\epsilon_m = 0.25$. The choice of these functions is motivated by the fact that the stiffness degrades over time in a periodic manner representing a possible fatigue crack growth in an aircraft over repeated pressurization. On the other hand, the mass increases and decreases over the nominal value due to re-fuelling and fuel burn over a flight period. The key consideration is that a digital twin of the dynamical system should track these types of changes by exploiting sensor data measured on the system.

7.2 OVERVIEW OF GAUSSIAN PROCESS EMULATORS

GP is a popular machine learning technique [18, 19] and has been successfully applied in the context of structural dynamic analysis [59] and finite element method [61]. Unlike other machine learning techniques such as the analysis-of-variance decomposition [39], support vector machine [87] and neural networks [56], GP adopts an optimal approach, infers a *distribution over functions* and then utilizes these functions to make predictions at some unknown points [134]. GP is also known as the Kriging [20, 166]. Different variants of GP exists in the literature. This includes fully Bayesian GP [19], sparse GP [5] and multi-fidelity GP [146] among other. In this chapter, *vanilla* GP is used. The discussion hereafter is focused on *vanilla* GP. For details on other types of GP, interested readers can consult [133].

Consider, $\ell \in \mathbb{R}^d$ as the input variable and

$$y = g(\ell) + \nu \tag{7.6}$$

to be a set of noisy measurements of the response variable, where ν represents the noise. The objective is to estimate the latent (unobserved) function $g(\ell)$ that enables prediction of the response variable, $\hat{y}$ at new values of ℓ. In GP-based regression, a GP is defined over $g(\ell)$ with the mean $\mu(\ell)$ and covariance function $\kappa(\ell, \ell'; \theta)$

$$g(\ell) \sim \mathcal{GP}\left(\mu(\ell), \kappa(\ell, \ell'; \theta)\right), \tag{7.7}$$

In the above equation

$$\begin{aligned}
\mu(\ell) &= \mathbb{E}\, g(\ell), \\
\kappa(\ell, \ell'; \theta) &= \mathbb{E}\left(g(\ell) - \mu(\ell)\right)\left(g(\ell') - \mu(\ell')\right).
\end{aligned} \tag{7.8}$$

and θ denotes the hyperparameters of the covariance function κ. The choice of the covariance function κ permits encoding of any prior knowledge about $g(\ell)$ (e.g., periodicity, linearity, smoothness), and can accommodate approximation of arbitrarily complex functions [208]. The notation in Equation (7.7) means that any finite collection of function values has a joint multivariate Gaussian distribution, that is $(g(\ell_1), g(\ell_2), \ldots, g(\ell_N)) \sim \mathcal{N}(\mu, \mathbf{K})$, where $\mu = [\mu(\ell_1), \ldots, \mu(\ell_N)]^T$ is the mean vector and $\mathbf{K}$ is the covariance matrix with $\mathbf{K}(i, j) = \kappa(\ell_i, \ell_j)$ for $i, j = 1, 2, \ldots, N$. If no prior information is available about the mean function, it is typically set to zero, i.e., $\mu(\ell) = 0$. However, for the covariance function, any function $\kappa(\ell, \ell')$ that generates a positive, semi-definite, covariance matrix $\mathbf{K}$ is a valid covariance function. Thus, the objective of GP is to estimate the hyperparameters, θ, based on the observed input-output pairs, $\{\ell_j, y_j\}_{j=1}^{N_t}$, where N_t is the number of training samples. Typically, this is achieved by maximizing the likelihood of the data [40]. Another approach is to adopt a Bayesian approach to compute the posterior of the hyperparameters, $\boldsymbol{\theta}$ [19]. Once $\boldsymbol{\theta}$ have been calculated, the predictive distribution of $g(\ell^*)$ given the dataset $\boldsymbol{\ell}, \boldsymbol{y}$, hyperparameters $\boldsymbol{\theta}$ and new inputs, ℓ^* is represented as

$$p(g(\ell^*)|\,\boldsymbol{y}, \boldsymbol{\ell}, \boldsymbol{\theta}, \ell^*) = \mathcal{N}\left(g(\ell^*)\mid \mu_{\mathcal{GP}}(\ell^*), \sigma_{\mathcal{GP}}^2(\ell^*)\right), \qquad (7.9)$$

where

$$
\begin{aligned}
\mu_{\mathcal{GP}}(\ell^*) &= \boldsymbol{k}^T(\boldsymbol{\ell}, \ell^*; \boldsymbol{\theta}) \left[\mathbf{K}(\boldsymbol{\ell}, \boldsymbol{\ell}; \boldsymbol{\theta}) + \sigma_n^2\,\mathbf{I}\right]^{-1}\boldsymbol{y}, \\
\sigma_{\mathcal{GP}}^2(\ell^*) &= k(\ell^*, \ell^*; \boldsymbol{\theta}) - \boldsymbol{k}^T(\boldsymbol{\ell}, \ell^*; \boldsymbol{\theta}) \left[\mathbf{K}(\boldsymbol{\ell}, \boldsymbol{\ell}; \boldsymbol{\theta}) + \sigma_n^2\,\mathbf{I}\right]^{-1}\boldsymbol{k}(\boldsymbol{\ell}, \ell^*; \boldsymbol{\theta}).
\end{aligned}
\qquad (7.10)
$$

For further details on GP, readers may refer [156]. Next, GP is used to obtain digital twin of a SDOF damped dynamic system.

7.3   GAUSSIAN PROCESS-BASED DIGITAL TWIN

The development of digital twin concept rests on the advances in the sensor technology. Using modern sensors, it is possible to collect different types of data such as acceleration time-history, displacement time-history, etc. In this chapter, we assume that using sensors, the natural frequency of the system described in Equation (7.1) is measured at some distinct time-steps. Advances in sensor and signal processing methods allow measurement of the frequencies of dynamic systems in real-time. In [221], a network-based real-time kinematic global navigation satellite system technique was adumbrated to monitor the displacements and vibration frequencies of a bridge. A wavelet packet filtering method was created to process experimental data. The method provides a potential technique for the dynamic monitoring of bridge structures. Liu et al. applied an extended Kalman filter to identify a structural system using acceleration measurements [122]. Gillich and Mituletu [83] performed spectral analysis on an acquired vibration signal to precisely estimate the natural frequencies of a structure in order to detect cracks. They tested their algorithm for real-world signals which included signals with damping and short signals. Sabato et al. [165] reviewed wireless micro electromechanical systems (MEMS) based acceleration sensor boards which can be applied for measuring low amplitude and low-frequency vibrations. They found that the technology for frequency measurement has advanced to the level where continuous monitoring of vibration of structures is now possible. Sony et al. [179] present non contact measurement systems such as smart sensing technology where smartphones, cameras, unmanned aerial vehicles (UAVs), and robotic sensors are used in acquiring and analyzing vibration data for improved structural condition monitoring. The development of such smart sensor technology has made the digital twin of dynamic systems-based on dynamic response a feasible task.

7.3.1 DIGITAL TWIN VIA STIFFNESS EVOLUTION

7.3.1.1 Formulation

In the first case, we assume that the change in the natural frequency with time is due to the deterioration of the stiffness of the system only. The equation of motion of the digital twin is now represented as

$$m_0 \frac{d^2 u(t)}{dt^2} + c_0 \frac{du(t)}{dt} + k(t_s) u(t) = f(t), \qquad (7.11)$$

where

$$k(t_s) = k_0 (1 + \Delta_k(t_s)). \qquad (7.12)$$

Equation (7.11) is a special case of Equation (7.1). In a practical scenario, generally no prior information exists about how the stiffness will deteriorate. It can be postulated that the deterioration of stiffness can be represented by using a GP

$$\Delta_k(t_s) \approx \Delta_{\hat{k}}(t_s) \sim \mathcal{GP}(\mu_k(t_s), \kappa_k(t_s, t'_s; \boldsymbol{\theta})). \qquad (7.13)$$

However, at this stage, there are no measurements corresponding to $\Delta_{\hat{k}}(t_s)$ and therefore, it is not feasible to estimate the hyperparameters, $\boldsymbol{\theta}$ of the GP. We substitute Equations (7.12) and (7.13) into Equation (7.11). Following this, the governing differential equation of the digital twin can be expressed as

$$m_0 \frac{d^2 u(t)}{dt^2} + c_0 \frac{du(t)}{dt} + k_0 \left(1 + \Delta_{\hat{k}}(t_s)\right) u(t) = f(t). \qquad (7.14)$$

Assuming a solution of the form $u(t) = \bar{u} \exp[\lambda t]$, substituting it in Equation (7.14) and solving the resulting quadratic equation for the free vibration (that is, $f(t) = 0$), the natural frequency of the system is obtained as

$$\lambda_{s_{1,2}}(t_s) = -\zeta_0 \omega_0 \pm i\omega_0 \sqrt{1 + \Delta_{\hat{k}}(t_s) - \zeta_0^2}, \qquad (7.15)$$

where ω_0 and ζ_0 are the natural frequency and damping ratio of the system at $t_s = 0$. Rearranging Equation (7.15) yields

$$\lambda_{s_{1,2}}(t_s) = -\underbrace{\frac{\zeta_0}{\sqrt{1 + \Delta_{\hat{k}}(t_s)}}}_{\zeta_s(t_s)} \underbrace{\omega_0 \sqrt{1 + \Delta_{\hat{k}}(t_s)}}_{\omega_s(t_s)} \qquad (7.16)$$

$$\pm \underbrace{i\omega_0 \sqrt{1 + \Delta_{\hat{k}}(t_s)} \sqrt{1 - \left(\frac{\zeta_0}{\sqrt{1 + \Delta_{\hat{k}}(t_s)}}\right)^2}}_{\omega_{d_s}(t_s)},$$

where $\omega_s(t_s) = \omega_0 \sqrt{1 + \Delta_{\hat{k}}(t_s)}$ is the evolution of the natural frequency, $\zeta(t_s) = \zeta_0 / \sqrt{1 + \Delta_{\hat{k}}(t_s)}$ is the evolution of the damping factor and

$\omega_{d_s}(t_s) = \omega_s(t_s)\sqrt{1 - \zeta_s^2(t_s)}$ is the evolution of the damped natural frequency with slower time-scale t_s. Most natural frequency extraction techniques extract the damped natural frequency of the system. We assume that the data obtained using the sensor to be the damped natural frequency. Following Ganguli and Adhikari [77], we can write

$$\Delta_{\hat{k}}(t_s) = -\tilde{d}_1(t_s)\left(2\sqrt{1 - \zeta_0^2} - \tilde{d}_1(t_s)\right), \tag{7.17}$$

where

$$\tilde{d}_1(t_s) = \frac{d_1(\omega_{d_0}, \omega_{d_s}(t_s))}{\omega_0}. \tag{7.18}$$

The function $d_1(\omega_{d_0}, \omega_{d_s}(t_s))$ in Equation (7.18) is the distance between ω_{d_0} and $\omega_{d_s}(t_s)$

$$d_1(\omega_{d_0}, \omega_{d_s}(t_s)) = ||\omega_{d_0} - \omega_{d_s}(t_s)||_2. \tag{7.19}$$

Since the initial damped frequency of the system, ω_{d_0} is known and sensor measurements for $\omega_{d_s}(t_s)$ exist, one can easily compute $\tilde{d}_1(t_s)$ and $\Delta_{\hat{k}}(t_s)$ by using Equation (7.18) and substituting it into Equation (7.17). Having said that, it is only reasonable to assume that the sensor measurements, $\omega_{d_s}(t_s)$, will be corrupted by some noise and therefore, the estimates of $\Delta_{\hat{k}}(t_s)$ are also noisy. Nevertheless, by considering these noisy estimates of $\Delta_{\hat{k}}(t_s)$ to be training outputs corresponding to the inputs t_s, the hyperparameters $(\boldsymbol{\theta})$ of the GP described in Equation (7.13) are obtained. In this chapter, we use Bayesian optimization [145] to maximize the likelihood of the data. Within the Bayesian optimization framework, L-BFGS optimizer is used with a tolerance of 10^{-5}. To prevent local convergence, multiple starting points were used. The hyperparameter $\boldsymbol{\theta}$ completely describes the digital-twin in Equation (7.14). An important query associated with GP is the form of the covariance function and the mean function. In this chapter, we have considered a pool for the mean function and the covariance function and then chosen the most suitable mean function and covariance function based on Bayesian information criteria. The possible candidates for the mean function and the covariance function are shown in Table 7.1. Functional forms of these covariance kernels are given in [208]. Details on Bayesian information criteria are provided next.

In this chapter, the optimal order of basis function and the optimal covariance kernel for GP are determined by using the Bayesian information criteria. The Bayesian information criteria corresponding to the m-th model is defined as [134]

$$bic_m = k_m \log(n) - \mathcal{L}\left(\hat{\theta}_m\right), \tag{7.20}$$

where k_m is the number of parameters and n represents the number data-point available to the m-th model. $\mathcal{L}(\boldsymbol{\theta})$ in Equation (7.20) represents the data-likelihood for the m-th model. The first term in Equation (7.20) penalizes a complex model, whereas the second term ensures that the model that best explains the model is selected. For GP, $\mathcal{L}(\boldsymbol{\theta}_m)$ follows a multi-variate Gaussian distribution [208].

Table 7.1

Candidate Functions for Mean and Covariance Functions in GP

Functions	Candidates
Mean function	Constant, Linear, Quadratic
Covariance function	Exponential, Squared Exponential, Matern 3/2, Matern 5/2, Rational Quadratic, ARD Exponential, ARD Squared Exponential, ARD Matern 3/2, ARD Matern 5/2, ARD Rational Quadratic

For determining the optimal model using the Bayesian information criterion, models corresponding to all possible combinations from mean function and covariance function is considered. The possible candidates for mean and covariance functions are shown in Table 7.1. There are 3 possible mean function and 10 possible covariance functions. Therefore, total 30 models were considered in this study. The BIC score is evaluated for each of the 30 models and the one with the minimum BIC score is selected.

7.3.1.2 Numerical Illustration

To apply the GP-based digital twin with only stiffness evolution, an SDOF system with nominal damping ratio $\zeta_0 = 0.05$ is considered. We assume that the sensor data is transmitted intermittently with a certain regular time interval. To simulate the variation in the natural frequency, we consider the change in the stiffness property of the system shown in Figure 7.1. and Figure 7.2(a) shows the actual change in the damped natural frequency of the system with time. The data-points available for the digital twin are also given in the figure. At this stage, it must be emphasized that the frequency of data availability depends on several factors such as the bandwidth of the wireless transmission system, cost of data collection, *etc.* In this case, we have considered 30 data-points. Figure 7.2(b) shows the GP-based digital twin model with clean data. It is observed that the GP-based digital twin captures the time evolution of Δ_k with great accuracy. However, note that this is an imaginary scenario as it is almost impossible to have measurements with no noise in any real system.

Now, consider a more realistic case where the data collected is contaminated by noise. For illustration purposes, a zero-mean Gaussian white noise with three different standard deviations is considered: (a) $\sigma_\theta = 0.005$, (b) $\sigma_\theta = 0.015$ and (c) $\sigma_\theta = 0.025$. Figures 7.3(a)–7.3(c) show the GP-based digital twin model corresponding to the three cases. For $\sigma_\theta = 0.005$, the GP-based digital twin successfully captures the evolution of stiffness with time. With

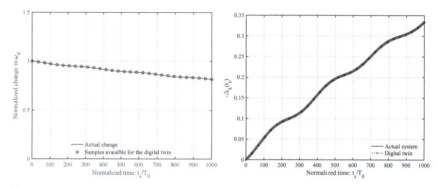

(a) Changes in the damped natural fre- (b) GP assisted digital twin with clean data
quency with time

Figure 7.2 Changes in the (damped) natural frequency and the GP-based digital twin obtained using the exact data plotted as a function of the normalized slow time t_s/T_0. Using Bayesian information criteria, "linear" based function and 'ARD Rational quadratic' covariance functions are selected. The estimated hyperparameters from Bayesian optimization are $\beta = [0.1804, 0.1009]$ and $\theta = [0.3729, 3.48 \times 10^8, 0.0128]$.

the increase in the noise level, a slight deterioration in the performance of the GP-based digital twin is observed. Nevertheless, even when $\sigma_\theta = 0.025$, the GP-based digital twin captures the time-evolution with great accuracy. An additional advantage of GP-based digital twin is its capability to capture the uncertainty emanating from limited data and noise. This is illustrated by a shaded plot in Figure 7.3. It can be observed that with increase in noise level, the uncertainty also increases.

To illustrate the benefits of having more data, a case with 100 data-points is shown in Figure 7.3(d). We observe that with the increase in the number of observations, the GP-based digital twin can track the evolution of stiffness in a more accurate manner even for the highest level of noise in the data.

7.3.2 DIGITAL TWIN VIA MASS EVOLUTION

7.3.2.1 Formulation

In the second case, consider that the time-evolution of the natural frequency is due to the change in the mass of the system. Subsequently, the equation of motion of the digital twin is represented as

$$m_s(t_s)\frac{d^2 u(t)}{dt^2} + c_0\frac{du(t)}{dt} + k_0 u(t) = f(t), \qquad (7.21)$$

where

$$m_s(t_s) = m_0(1 + \Delta_m(t_s)). \qquad (7.22)$$

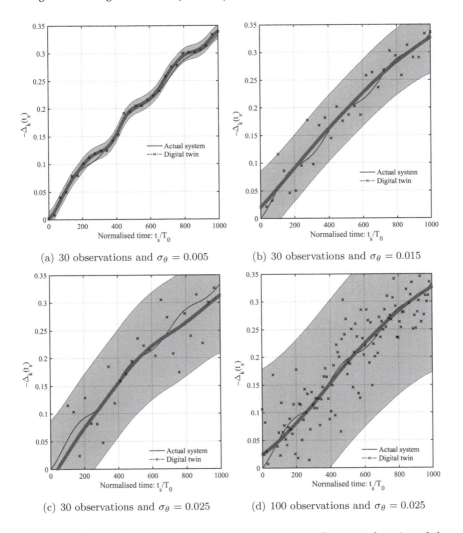

(a) 30 observations and $\sigma_\theta = 0.005$

(b) 30 observations and $\sigma_\theta = 0.015$

(c) 30 observations and $\sigma_\theta = 0.025$

(d) 100 observations and $\sigma_\theta = 0.025$

Figure 7.3 GP-based digital twin constructed with noisy data as a function of the normalized slow time t_s/T_0. For (a) – (c), Bayesian information criteria selects "constant" basis function. The "ARD Rational quadratic," "ARD squared exponential," and "ARD Matern 3/2" covariance kernels are selected for cases (a) to (c), respectively. For (d), Bayesian information criteria yields "Constant" basis with $\beta = 0.1768$ and "ARD Matern 3/2" covariance kernel with $\boldsymbol{\theta} = [2.7298 \times 10^3, 0.3402]$. The shaded graph depicts the 95% confidence interval.

Similar to case 1, assume

$$\Delta_m \left(t_s \right) \approx \Delta_{\hat{m}} \left(t_s \right) \sim \mathcal{GP} \left(\mu_m \left(t_s \right), \kappa_{\hat{m}} \left(t_s \right) \right), \tag{7.23}$$

substitute Equations (7.22) and (7.23) into Equation (7.21) and solve it to get the natural frequency of the system

$$\lambda_{s_{1,2}}(t_s) = -\omega_s(t_s)\zeta_s(t_s) \pm i\omega_{d_s}(t_s), \tag{7.24}$$

where

$$\omega_s(t_s) = \frac{\omega_0}{\sqrt{1 + \Delta_{\hat{m}}(t_s)}}, \tag{7.25a}$$

$$\zeta_s(t_s) = \frac{\zeta_0}{\sqrt{1 + \Delta_{\hat{m}}(t_s)}} \quad \text{and} \tag{7.25b}$$

$$\omega_{d_s}(t_s) = \omega_s(t_s)\sqrt{1 - \zeta_s^2(t_s)}. \tag{7.25c}$$

represent the evolution of natural frequency, damping ratio and damped natural frequency of the digital twin. Again, following a procedure similar to case 1, we get

$$
\begin{aligned}
\Delta_{\hat{m}}(t_s) =& \frac{-2\tilde{d}_2(t_s)^2 + 4\tilde{d}_2(t_s)\sqrt{1 - \zeta_0^2} - 1 + 2\zeta_0^2}{2\left(-\tilde{d}_2(t_s) + \sqrt{1 - \zeta_0^2}\right)^2} \\
&+ \frac{\sqrt{1 - 4\tilde{d}_2(t_s)^2\zeta_0^2 + 8\tilde{d}_2(t_s)\sqrt{1 - \zeta_0^2}\zeta_0^2 - 4\zeta_0^2 + 4\zeta_0^4}}{2\left(-\tilde{d}_2(t_s) + \sqrt{1 - \zeta_0^2}\right)^2}
\end{aligned}
\tag{7.26}
$$

where $\tilde{d}_2(t_s)$ is the equivalent of $\tilde{d}_1$ for the mass evolution case. Similar to case 1, $\Delta_{\hat{m}}(t_s)$ obtained using Equation (7.26) will be noisy in nature. We consider this noisy estimates of $\Delta_{\hat{m}}(t_s)$ to be the training outputs and t_s to be the training inputs for estimating the hyperparameters, $\boldsymbol{\theta}$, of the GP defined in Equation (7.23). This is achieved by maximizing the likelihood of the data. The parameter settings for solving the optimization problem are kept similar as before. For determining the best mean function and covariance function, Bayesian information criterion mentioned before is again adopted. Once the hyperparameter, $\boldsymbol{\theta}$, is estimated, the digital twin model defined in Equation (7.21) is completely defined.

7.3.2.2 Numerical Illustration

To illustrate the applicability of the GP-based digital twin model for mass evolution, we revisit the SDOF example considered for stiffness evolution. To simulate the variation in the natural frequency, consider the change in the mass of the system shown in Figure 7.1. Here Figure 7.4(a) shows the actual change in the damped natural frequency of the system with time. The datapoints available for the digital twin are also shown in the figure. Note that the evolution of mass with time is more complex, and therefore, unlike for the stiffness evolution case, using only 30 observations is inadequate. Therefore, we have assumed access to more data points in this case. Figure 7.4(b) shows the GP-based digital twin developed from clean data. The GP-based digital twin can capture the time evolution of mass with a high degree of accuracy. Furthermore, the uncertainty due to limited data is captured adequately. However, as stated earlier, this zero-noise case is an unrealistic case as in a practical scenario, data collected will always be corrupted by some form of noise.

Next, consider a more realistic case where the data collected is polluted by noise. Figure 7.5 shows the GP-based digital twin model trained with 100 noisy observations and $\sigma_\theta = 0.005$. Some discrepancy between the GP-based digital twin and the physical twin is observed. For better performance, the number of observations is increased to 150. The GP-based digital twin corresponding to the three noise levels are shown in Figure 7.6(a)–(c). In this case, the GP-based digital twin yields accurate result. Lastly, Figure 7.7 shows the GP-based digital twin with 200 samples and $\sigma_\theta = 0.025$. This case yields the best result with accurate time evolution of mass and uncertainty quantification. Overall, this example illustrates the importance of (a) a higher sampling rate and (b) denoising of data, for digital twin technology.

7.3.3 DIGITAL TWIN VIA MASS AND STIFFNESS EVOLUTION

7.3.3.1 Formulation

As the last case, consider the simultaneous evolution of the mass and stiffness. The governing differential equation for this case is given as

$$m_s(t_s)\frac{\mathrm{d}^2 u(t)}{\mathrm{d}t^2} + c_0\frac{\mathrm{d}u(t)}{\mathrm{d}t} + k_s(t_s)u(t) = f(t), \tag{7.27}$$

where $k_s(t_s)$ and $m_s(t_s)$ are given by Equations (7.12) and (7.22), respectively. We express $\Delta_m(t_s)$ and $\Delta_k(t_s)$ by a multi-output GP [5, 19],

$$[\Delta_m(t_s), \Delta_k(t_s)] \approx [\Delta_{\hat{m}}(t_s), \Delta_{\hat{k}}(t_s)] \sim \mathcal{GP}(\boldsymbol{\mu}_{t_s}, \boldsymbol{\kappa}(t_s, t_s'; \boldsymbol{\theta})), \tag{7.28}$$

Substituting Equations (7.28), (7.12), and (7.22) in 7.27 and solving, we get

$$\lambda_{s_{1,2}} = -\omega_s(t_s)\zeta_s(t_s) \pm \mathrm{i}\omega_{d_s}(t_s), \tag{7.29}$$

where

$$\omega_s(t_s) = \omega_0\frac{\sqrt{1 + \Delta_{\hat{k}}(t_s)}}{\sqrt{1 + \Delta_{\hat{m}}(t_s)}} \tag{7.30a}$$

$$\zeta_s(t_s) = \frac{\zeta_0}{\sqrt{1 + \Delta_{\hat{m}}(t_s)}\sqrt{1 + \Delta_{\hat{k}}(t_s)}} \quad \text{and} \tag{7.30b}$$

$$\omega_{d_s}(t_s) = \omega_s(t_s)\sqrt{1 - \zeta_s^2}. \tag{7.30c}$$

Note that unlike for the previous two cases, it is not possible to calculate the hyperparameters of the GPs-based on measurements of damped natural

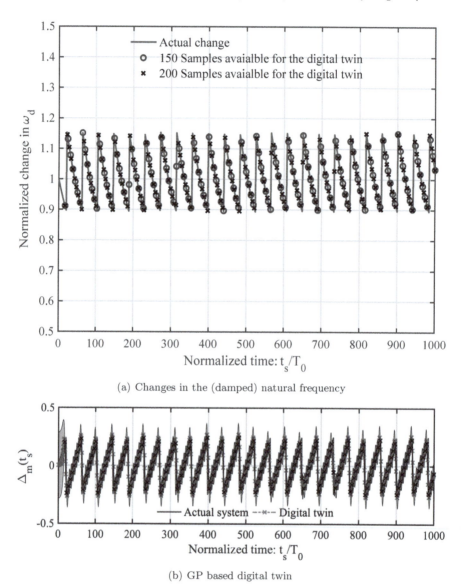

(a) Changes in the (damped) natural frequency

(b) GP based digital twin

Figure 7.4 Changes in the (damped) natural frequency and the GP-based digital twin obtained using the exact data plotted as a function of the normalized slow time t_s/T_0. Using Bayesian information criteria, "Constant" basis function and "ARD Matern 5/2" covariance functions are selected. The estimated hyperparameters from Bayesian optimization are $\beta = 0.0$ and $\boldsymbol{\theta} = [4.1532, 0.1429]$. The shaded plot depicts the 95% confidence interval.

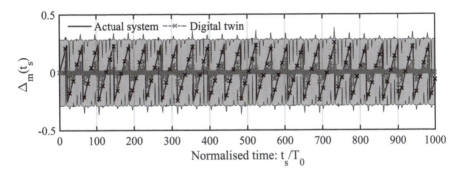

Figure 7.5 GP-based digital twin obtained using 100 noisy data with $\sigma_\theta = 0.005$, plotted as a function of the normalized slow time t_s/T_0. The shaded plot depicts the 95% confidence interval.

frequencies only. Therefore, we take a different path by considering the real and imaginary part in Equation (7.29) separately. Following results derived in Chapter 5, it can be shown that

$$\Delta_{\hat{m}}(t_s) = -\frac{\tilde{d}_\mathcal{R}(t_s)}{\zeta_0 + \tilde{d}_\mathcal{R}(t_s)}, \tag{7.31a}$$

$$\Delta_{\hat{k}}(t_s) = \frac{\zeta_0 \tilde{d}_\mathcal{R}^2(t_s) - \left(1 - 2\zeta_0^2\right)\tilde{d}_\mathcal{I}(t_s) + \zeta_0^2 \tilde{d}_\mathcal{I}^2(t_s)}{\zeta_0 + \tilde{d}_\mathcal{R}(t_s)}, \tag{7.31b}$$

where $\tilde{d}_\mathcal{R}(t_s)$ and $\tilde{d}_\mathcal{I}(t_s)$, as before, are distance measures

$$\tilde{d}_\mathcal{R}(t_s) = \frac{d_\mathcal{R}(t_s)}{1 + \Delta_{\hat{m}}(t_s)} \tag{7.32}$$

$$\tilde{d}_\mathcal{I}(t_s) = \sqrt{1 - \zeta_0^2} - \frac{\sqrt{\left(1 + \Delta_{\hat{k}}(t_s)\right)\left(1 + \Delta_{\hat{m}}(t_s)\right) - \zeta_0^2}}{1 + \Delta_{\hat{m}}(t_s)}.$$

Note that Equation (7.31) provides estimates $\Delta_{\hat{m}}(t_s)$ and $\Delta_{\hat{k}}(t_s)$ based on noisy measurements, and therefore, are also noisy. Using these noisy estimates as training outputs and t_s as the training inputs, we estimate the hyperparameters, $\boldsymbol{\theta}$ of the multi-output GP given in Equation (7.28). This is accomplished by maximizing the likelihood of the data as described in Equation (7.2). The same parameter settings as adumbrated before are considered. To find the best mean function and covariance function, Bayesian information criteria is adopted as earlier in this chapter. The hyperparameters, $\boldsymbol{\theta}$ are estimated and then completely defines the digital twin described in Equation (7.27).

7.3.3.2 Numerical Illustration

We revisit the previously investigated SDOF system to illustrate the applicability of GP-based digital twin for simultaneous mass and stiffness evolution.

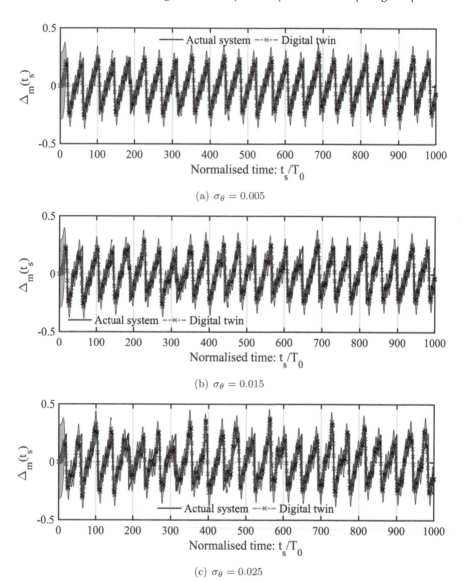

(a) $\sigma_\theta = 0.005$

(b) $\sigma_\theta = 0.015$

(c) $\sigma_\theta = 0.025$

Figure 7.6 GP based digital twin obtained using 150 noisy data plotted as a function of the normalized slow time t_s/T_0. For (a)–(c), Bayesian information criteria selects "Constant" basis and "ARD Matern 5/2" covariance kernel. The shaded plot depicts the 95% confidence interval.

To simulate the variation in the natural frequency, we use the change in the mass of the system shown in Equation (7.1). Figure 7.8 shows the actual change in real and imaginary parts of the natural frequency of the system

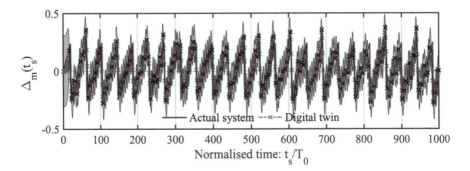

Figure 7.7 GP-based digital twin obtained using 200 noisy data with $\sigma_\theta = 0.025$ plotted as a function of the normalized slow time t_s/T_0. Bayesian information criteria yields "Constant" basis with $\beta = 0.0$ and "ARD Matern 5/2" covariance kernel with $\boldsymbol{\theta} = [4.1242, 0.1527]$.

over time. Similar to the previous cases, the damping ratio is assumed to be 0.05. Figure 7.9 shows the GP-based digital twin constructed from clean data. The trained GP-based digital twin can capture the time evolution of mass and stiffness well. However, data with no noise is an unrealistic scenario as despite the use of the most advanced sensors, the data collected will be noisy due to the environmental effects on a system [223].

Now, consider a more realistic case where the sensor data is laced with noise. Similar to the previous two cases, we consider three noise levels. Figure 7.10 shows the mass and stiffness evolution of the GP-based digital twin trained with 37 noisy observations. The observations are laced with white

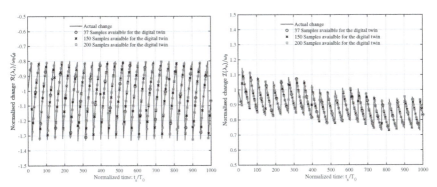

(a) Changes in real part of natural frequency

(b) Changes in imaginary part of natural frequency

Figure 7.8 Normalized changes in the real and imaginary parts of the natural frequency as a function of the normalized slow time t_s/T_0. The shaded plot depicts the 95% confidence interval.

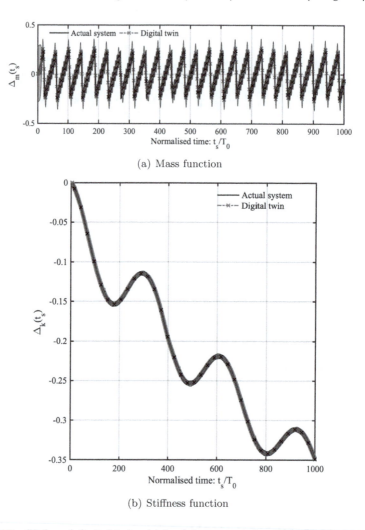

(a) Mass function

(b) Stiffness function

Figure 7.9 GP-based digital twin obtained from exact data via simultaneous mass and stiffness evolution as a function of the normalized 'slow time t_s/T_0. Bayesian information criteria yields "Linear" basis and "squared exponential" covariance kernel.

Gaussian noise having $\sigma_\theta = 0.005$. It is found that the GP-based digital twin can capture the time evolution of stiffness with high accuracy. However, the GP-based digital twin fails to capture the mass evolution in an appropriate manner. This is expected as the mass evolution function used in this chapter is quite complex, and it is unreasonable to think that with such meager data, GP will be able to track the mass evolution. Note that the uncertainty caused by limited and noisy data is perfectly captured, and therefore, the true solution resides within the shaded portion.

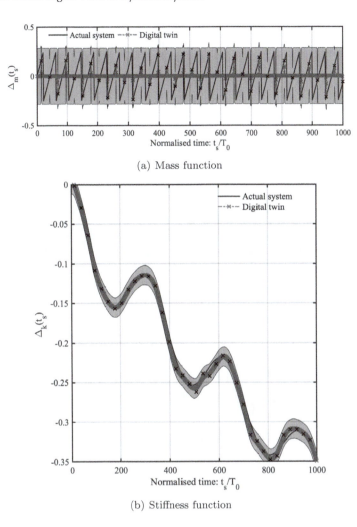

(a) Mass function

(b) Stiffness function

Figure 7.10 GP-based digital twin obtained from 37 noisy data with $\sigma_\theta = 0.005$ via simultaneous mass and stiffness evolution as a function of the normalized slow time t_s/T_0. Bayesian information criteria yields "Linear" basis and "squared exponential" covariance kernel. The shaded plot depicts the 95% confidence interval.

Figures 7.11 and 7.12 show the evolution of mass and stiffness of the GP-based digital twin trained with 150 noisy samples. For mass evolution, results corresponding to all three noise levels are shown. With the increase in the number of samples, a dramatic improvement in the GP-assisted digital twin is seen. The evolution of stiffness is shown only for $\sigma_\theta = 0.025$ and results obtained are very accurate. Lastly, evolution of mass with 200 noisy observations and $\sigma_\theta = 0.025$ is shown in Figure 7.13. As expected, this case yields the best results.

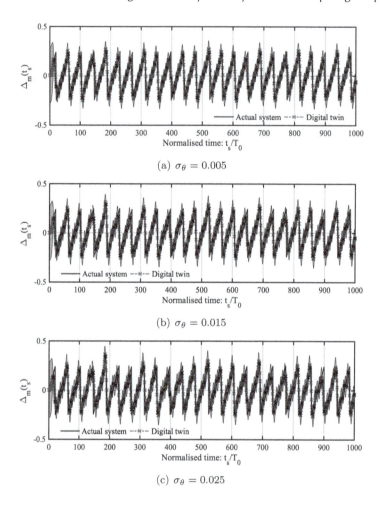

(a) $\sigma_\theta = 0.005$

(b) $\sigma_\theta = 0.015$

(c) $\sigma_\theta = 0.025$

Figure 7.11 Mass evolution as a function of the normalized slow time t_s/T_0 for GP-based digital twin (simultaneous mass and stiffness evolution) obtained from 150 noisy data. Noise levels of $\sigma_\theta = 0.005$, $\sigma_\theta = 0.015$ and $\sigma_\theta = 0.025$ are considered. Bayesian information criteria yields "Linear" basis and "ARD Matern 5/2" covariance kernel. The shaded plot depicts the 95% confidence interval.

7.4 DISCUSSION

Although in theory, a digital twin of a physical system can be achieved in several ways, most of the existing works on digital twin have focused on the broader conceptual aspect. In this chapter, we follow a different path and focus on a specific case of a structural dynamical system. More specifically, we have applied surrogate models such as GP with a SDOF system. The key ideas developed in this chapter include:

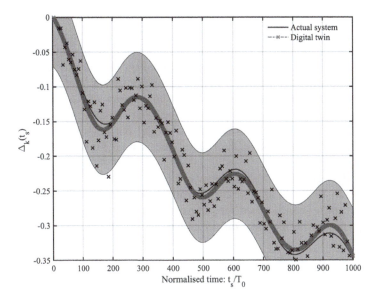

Figure 7.12 Stiffness evolution as a function of the normalized slow time t_s/T_0 for GP-based digital twin (simultaneous mass and stiffness evolution) trained with 150 noisy data with $\sigma_\theta = 0.025$. Bayesian information criteria yields "Linear" basis and "ARD Matern 5/2" covariance kernel. The shaded plot depicts the 95% confidence interval.

The concept of a surrogate model within the digital twin technology is introduced. The use of GP within the digital twin technology is advocated. The GP is trained based on observations on a slow timescale t_s and then used to predict the model parameters in the fast time-scale t.

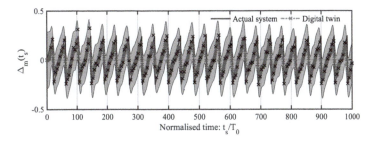

Figure 7.13 Mass evolution as a function of the normalized slow time t_s/T_0 for GP based digital twin (simultaneous mass and stiffness evolution) trained with 200 noisy data with $\sigma_\theta = 0.025$. Bayesian information criteria yields "Linear" basis and "ARD Matern 5/2" covariance kernel. The shaded plot depicts the 95% confidence interval.

The importance of collecting more data during the system life-cycle is illustrated in this chapter through several numerical examples. A dramatic improvement in the performance of the digital twin is observed with an increase in the number of data collected. In other words, over the same window of time, a higher sampling rate is preferable.

The importance of cleaner data in the context of a digital twin is illustrated in this chapter. With less noise in the data, the digital twin can track the time evolution of the model parameters with alacrity. Digital signal processing algorithms can be used for data cleaning [74, 162].

Since GP is a Bayesian surrogate model, it can quantify the uncertainty in the system emanating from limited data and noise. These uncertainties can be used for judging if there is a need to increase the frequency of data collection, i.e., increase the measurement sampling rate.

Overall, GP is able to capture both mass and stiffness variation from limited noisy data; although, for capturing the time evolution of mass, more observations are needed as compared to the stiffness evolution case.

The system studied in this chapter is a simple SDOF dynamical system governed by second-order ordinary differential equations. The same framework applies to other physical systems (e.g., simple electrical systems involving a resistor, capacitor, and an inductor) governed by this kind of equations. In principle, it is possible to expand the proposed framework to discrete MDOF systems or discretized continuum systems with proportional damping. This will require new derivations regarding the physics-informed part of the digital-twin framework. This has been touched upon in Chapter 9. The framework presented here can be expanded for more rigorous investigations encompassing a wider variety of practical problems. Some of the future possibilities are highlighted below:

1. *Surrogate-assisted digital twin for big data:* In this chapter, natural frequency measurements are applied for establishing the GP-based digital twin. However, it will be more useful if the GP-based digital twin can be trained directly from the time-history of measured responses. This is essentially a big-data problem and *vanilla* GP is unable to address such problems. More advanced versions of GP, such as the sparse GP [177] and convolution GP [6] can be explored in that case.

2. *Surrogate-based digital twin for continuum systems:* For many engineering problems, a continuum model represented by the partial differential equation is the preferred physics-based modeling choice. Developing a surrogate-based digital twin for such a system will be useful. This is primarily a sparse data scenario as sensor responses will only be available at

few spatial locations. Methods such as convolution neural networks [151] can be used for such systems.

3. *Digital twin for systems with unknown/imperfect physics:* In literature, there exists a plethora of problems for which the governing physical law is not well defined. It is interesting and extremely useful to learn/develop the digital twin purely based on data. Works on discovering physics from data by using machine learning can be found in the literature [216]. Genetic programming can also be used to fit mathematical functions to data [175].

4. *Surrogate-based digital twin models with machine learning:* Over the past few years, the machine learning community has witnessed brisk progress with developments of techniques such as deep neural networks [27, 28, 85], convolution neural networks [151], among others. Using these techniques within the digital twin framework can elevate this technology to new heights. The required sampling rate and data quality could be reduced by using machine learning methods.

5. *Multiple-degrees-of-freedom (MDOF) digital twins using surrogate models:* In this chapter, we focused on a SDOF model, which can be considered to be a simple idealization of complex MDOF systems. However, for more effective digital twin with realistic predictive capabilities, MDOF digital twin should be considered. To that end, it is necessary to develop surrogate-based MDOF digital twins.

6. *Surrogate models for nonlinear digital twins:* The model considered in this chapter is a linear ordinary differential equation. However, many physical systems show nonlinear behavior. A classic example of a nonlinear system is the Duffing oscillator [136]. For such a nonlinear system, it is unlikely that a closed-form solution will exist and hence, developing a surrogate-based digital twin model will be useful.

7. *Forecasting using surrogate models for digital twins:* One of the primary tasks of a digital twin is to provide future prediction. The physics-based digital twin proposed previously (Chapter 5) is not amenable for prediction of the future response. The GP-based digital twin model, on the contrary, can predict the short-term future response. It is necessary to develop surrogate-based digital twin capable of forecasting the long-term response of the system.

8. *Predicting extremely low-probability catastrophic events for digital twins with surrogate models:* Till date, application of digital twin technology is mostly limited to maintenance, prognosis and health monitoring. Other possible applications of digital twin involve computing probability of failure, rare event probability and extreme events.

9. *High-dimensional surrogate models for multi time-scale digital twins:* In this chapter, we assumed one time-scale for the evolution of the digital twin. However, there is no physical or mathematical reason as to why this must be restricted to only one time-scale. It is possible that the parameters may evolve at different time-scales. Such systems are known as multi-scale

dynamical system [42]. Developing surrogate-based digital twin for such multi-scale dynamical systems needs further investigation. We will address this issue in the next chapter.

10. *Hybrid surrogate models for digital twins:* In this chapter, only one type of surrogate model, namely, the Gaussian process emulator is used. Several types of surrogate models have been considered to solve a wide range of complex problems with different numbers of variables [40, 175]. It is well known that some surrogate models perform better than the others in given situations. It is plausible to employ multiple surrogate models simultaneously by taking the benefit of their relative strengths. Such hybrid surrogate digital twins can exhibit superior performance compared to single surrogate digital twins.

7.5 SUMMARY

Surrogate models have been developed over the past several decades with increasing computational efficiency, sophistication, variety, depth and breadth. They are a class of machine learning methods which benefit from the availability of data and superior computing power. As digital twins are also expected to exploit data and computational methods, there is a compelling case for the use of surrogate models in this context. Motivated by this synergy, this chapter explored the possibility of using a particular surrogate model, namely, the GP emulator, for the digital twin of a damped SDOF dynamic system. The proposed digital twin evolves at a time-scale which is much slower than the dynamics of the system. This makes it possible to identify crucial system parameters as a function of the "slow-time" from continuously measured data. Closed-form expressions derived considering the dynamics of the system in the fast time-scale have been used. The Gaussian process emulator is employed an the slow time-scale, where the impact of lack of data (sparse data) and noise in the data have been investigated.

The results obtained by applying the GP emulator show that sparsity of data may preclude the GP from accurately capturing the evolution of mass and stiffness. This is obvious for the mass evolution case where a higher sampling rate is needed to accurately capture the temporal evolution of mass. Uncertainty in the GP-based digital twin model results from both sparsity in the data and noise in the measurements. Therefore, only increasing the number of observations may not ameliorate the uncertainty in the system. On the other hand, collecting clean data (i.e., data with lower noise level) allows the GP to track the evolution of mass and stiffness evolution more accurately. Nevertheless, even for the cases with a relatively larger noise level, GP yields an accurate result. Moreover, GP also captures the uncertainty due to limited and sparse data.

Although only the GP emulator with SDOF dynamic system is considered in this chapter, several conceptual extensions directly follow from this work. The approach proposed here exploits closed-form expressions based on the physics of the system and a surrogate model for the measured data. However, the proposed surrogate model-based approach is eminently suitable when more advanced computational model is used for the physical problem (for example, when a MDOF system is considered). Therefore, the underlying formulation is not limited to the SDOF example. The overall framework suggests a paradigm for a hybrid physics-based and data-driven digital twin approach. The physics-based approach is linked to the concept of "fast time," while the data-driven approach is operational on the "slow time" domain. The separation of scientific approaches based on the inherently different timescales permits the integration of multi-disciplinary approaches in the future development of the digital twin technology.

8 Digital Twin at Multiple Time Scales

Design and analysis of complex engineering systems using high fidelity computational simulations are a key aspect of modern engineering practice. In the context of aerospace and mechanical engineering, computer models and simulations are often employed to support conceptual design, prototyping, manufacturing, production, test-data correlation and safety assessment. Over the last decade, there is a shift toward exploiting computer simulations for providing service throughout the whole product life cycle [154, 183], going well beyond the production stage. In the context of civil infrastructure, the concept of fusing digital information with real-life structures are also evolving with alacrity [7]. The methodologies, algorithms, techniques, software and computer applications which mimic the evolution of a complex real system through computational and digital means are broadly named as "digital twins." The main technologies that enable the espousal of digital twins include Artificial Intelligence (AI)/Machine Learning (ML), and Internet-of-Things (IoT), among others. Furthermore, factors such as the growing usage of connected devices across organizations, increasing adoption of cloud platforms, and the emergence of high-speed networking technologies further fuels the growth of the digital twin technology.

By its very definition, digital twins are extremely diverse resulting in very different approaches to different applications. In this chapter, we are interested in digital twins of structural dynamic systems. A digital twin is a virtualized proxy of a real physical dynamic system. While a numerical model of a physical system seeks to closely match the behavior of a dynamic system, the digital twin also tracks the temporal evolution of the dynamic system. Once a digital twin is trained and developed, it can be employed to make crucial decisions at a point of time which is significantly far in the future from the time of manufacturing of an engineering dynamic system. A general mathematical framework for digital twin is suggested in [210]. More specific approaches to developing digital twins include prognostics and health monitoring [21, 88, 124, 182, 191], manufacturing [58, 89, 91, 123, 143], automotive and aerospace engineering [92, 100, 187, 218], to mention a few. These references give an excellent idea of what can be accomplished using the digital twin approach.

Philosophically, the development of digital twins can either be considered from a physics-based modeling point of view or from a data-based modeling point of view. While physics-based approaches are often more robust, there are cases where physics of a given system is unknown or only partially known. In other cases, the physics of the problem may require the solution of onerous mathematics. Under such circumstances, it is not possible to use a purely physics-based modeling approach. Naturally, data-based modeling approaches emerge as the only viable alternative. The exact application of data-based modeling approaches is a function of on the underlying problem. In case the physics of the problem is completely unknown, the problem can be pursued from a purely data-driven point-of-view. On the other hand, if the physics of the problem is partially known, data-driven modeling can be used to compensate for the missing physics. Historically, physics-based approaches have been ubiquitous for the dynamic analysis of complex structures. However, with the development of sensor technology and wireless data transmission, the availability of useful data is increasing. Keeping this in mind, a hybrid physics-based and data-based approach for digital twin is pursued in this chapter, which builds on the theory developed in Chapter 8.

Dynamic systems differ from other systems because of the fact that their response due to external excitation changes with time. The rate of change of response depends on their characteristics time period. This is a fundamental property of a dynamic system and depends on the stiffness to mass ratio of the system. Typically, smaller structures have faster time periods and larger structures, for example, a large wind turbine, have smaller time periods. However, irrespective of their characteristics time periods, the time-scale of their operational life is considerable. As an example, for a large wind turbine, the time period is in the order of tens of seconds [3], while its operation life is in tens of years. To account for this fundamental mismatch in the time-scales, the digital twin proposed in Chapter 5 explicitly considers two different time scales. The intrinsic timescale is a fast time scale, while the operational time scale is a slow time scale. Physics-based methods, such as the finite element method, are used for dynamic evolution in the intrinsic timescale. Data-based methods (e.g., surrogate models) are used for dynamic evolution in the operational timescale. The digital twin of a complex dynamic system rises from the fusion of physics and data-based approaches, also known as the gray box modeling technique. The separation of computational approaches based on the two different time scales was exploited in Chapter 7 where Gaussian Process Emulators (GPE) was used in the slow time-scale.

An important feature of digital twin technology is to use the sensor data collected from the physical system to update the digital twin and also to predict the future states. In this regard, the role of ML algorithms become enormous. One of the reasons behind the recent progress in digital twin technology is the development of advanced ML algorithms (e.g., deep neural network [27, 28, 85, 168], Gaussian process [18, 19, 135]) that can be used to update the model and make predictions. For example, [21] applied two deep learning algorithms within the digital twin framework for prognosis and diagnosis of systems. Similarly, in [30], GPE was applied to learn the evolution of the system parameters. Forecasting models such as autoregressive moving average [167] and autoregressive integrated moving average [23] are also useful. A detailed review of the impact of ML algorithms on the digital twin technology is given by Kaur et al [103].

The separation of two time-scales provides a logical framework for developing computational and mathematical methods for the construction of digital twins. However, questions remain on how to define and apply the slow time scale. The fast time-scale t is a fundamental property of dynamical systems and therefore is unambiguous for a given system. The same is not true for the slow operational time scale t_s. In Chapters 5 and 7, the idea of a single time-scale for the evolution of the digital twin was used for its entire operational period. However, there is no physical or mathematical reason as to why the evolution of the digital twin must be restricted to only one time-scale. It is plausible that various parameters in a complex digital twin evolve at different time scales. For example, the mass of a system can change due to corrosion, while the stiffness of a system can degrade due to fatigue. These two processes will have disparate time scales of evolution. Therefore, a digital twin can evolve in different time scales in addition to its intrinsic time scale. The solution of such systems is onerous from a computational point-of-view as we need a time-step of the order of the fastest scale. Therefore, tasks such as health-monitoring, damage prognosis and remaining useful-life prediction becomes almost intractable. The difficulty increases dramatically for cases where the underlying physics governing the time-evolution is unknown and data available is noisy. The key idea proposed and investigated in this chapter is that the digital twin of a dynamic system evolves in two different operational time-scales. In principle, there can be more than two operational time-scales. Approaches proposed in the paper can form the basis of considering such problems.

The rest of the chapter is organized as follows. In Section 8.1, the problem undertaken in this chapter is discussed. Details about the proposed digital twin framework for the multi-timescale dynamical system are provided in Section 8.2. The performance of the digital twin in capturing the (multiscale) temporal evolution of the system parameters is provided in Section 8.3. Finally, Section 8.4 presents the summary of this chapter. This chapter is adapted from Ref. [29].

8.1 THE PROBLEM STATEMENT

Again, consider a physical system that can be represented by a single-degree-of-freedom (SDOF) spring mass and damper system.

$$m_0 \frac{\mathrm{d}^2 u_0(t)}{\mathrm{d}t^2} + c_0 \frac{\mathrm{d}u_0(t)}{\mathrm{d}t} + k_0 u_0(t) = f_0(t), \qquad (8.1)$$

where m_0, c_0 and k_0 are, respectively the mass, damping and stiffness of the system. Here t is the intrinsic time of the system. Equation (8.1) is often referred to as the "nominal system" and m_0, c_0, and k_0 as the nominal mass, nominal damping and nominal stiffness, respectively. $f_0(t)$ and $u_0(t)$ are respectively the forcing function and the dynamic response of the nominal system. Note that a more realistic infinite-dimensional system expressed by using partial differential equations can be discretized into finite-dimensional systems by using standard numerical techniques such as the Galerkin method. These discretized systems are often condensed into SDOF systems (as in Equation 8.1) using orthogonal transformations.

The nominal system discussed in Equation (8.1) has fixed system parameters m_0, c_0, and k_0. For a digital twin, however, the system parameters, namely mass, damping and stiffness, and the forcing function changes with the service time t_s. A generalized equation of motion of this system can be represented as

$$m(t_s)\frac{\partial^2 u(t, t_s)}{\partial t^2} + c(t_s)\frac{\partial u(t, t_s)}{\partial t} + k(t_s)u(t, t_s) = f(t, t_s). \qquad (8.2)$$

Typically, the evolution of parameters with service time t_s is slow. The nominal system discussed in Equation (8.1) can be viewed as the initial model at $t_s = 0$. The service time t_s can represent the number of cycles in a aircraft. From Equation (8.2), we note that the mass $m(t_s)$, damping $c(t_s)$, stiffness $k(t_s)$, and $f(t, t_s)$ changes with the 'service time' t_s, for instance due to the degradation in the system during its service time. Equation (8.2) represents the the equation of motion of the digital twin. Note that when $t_s = 0$, Equation (8.2) devolves to the nominal system represented in Equation (8.1). It is clear that the digital twin is completely described by the functions $m(t_s)$, $c(t_s)$, and $k(t_s)$. Therefore, for using the digital twin in the field, the functions $m(t_s)$, $c(t_s)$, and $k(t_s)$ must be obtained.

In earliest chapters, physics-based and data-based approaches for estimating the functions $m(t_s)$, $c(t_s)$ and $k(t_s)$ have been developed. However, these studies showed a number of limitations.

> The physics-based digital twin proposed in Chapter 5 is insufficiently accurate when the sensor data is noisy.
> The data-based digital twin proposed in Chapter 7 only works for systems with a single operational time-scale. The approach is inapplicable for multi-timescale dynamical systems [42, 106].
> One of the objectives of digital twin is to predict the future response, so as to understand the behavior of the physical twin in future. Unfortunately, neither the physics-based nor the data-based digital twins previously proposed are capable of predicting the future responses.

The objective of this chapter is to develop an efficient framework for resolving some of the above-mentioned limitations. More specifically, we want to develop digital twins for multi-timescale dynamical systems. Unlike the digital twins developed in Chapters 5 and 7, the digital twin developed in this chapter is able to predict future responses.

To develop the digital twin, it is assumed that sensors are deployed on the physical system. Recent developments in the field of IoT provides us with numerous new data collection technologies. This provides the necessary connectivity between the physical and digital twins. Using the sensors, measurements are taken intermittently at t_s. It is assumed that the functions $m(t_s)$, $c(t_s)$, and $k(t_s)$ are so slow that the dynamics of the system in Equation (8.2) is decoupled. In other words, m_s, c_s, and k_s of the system is constant relative the instantaneous dynamics of the system. Without loss of generality, we assume

$$k_s(t_s) = k_s \left(t^{(s)}, t^{(f)} \right) = k_0 \left(1 + \Delta_{\tilde{k}} \left(t^{(s)}, t^{(f)} \right) \right) \tag{8.3}$$

where

$$1 + \Delta_{\tilde{k}} \left(t^{(s)}, t^{(f)} \right) = \Delta_k^{(s)} \left(t^{(s)} \right) + \Delta_k^{(f)} \left(t^{(f)} \right) = \Delta_k \left(t^{(s)}, t^{(f)} \right). \tag{8.4}$$

Here $t^{(s)}$ and $t^{(f)}$ represent a slower and a faster time scale of evolution of the respective processes. The "multi-scale" nature of the problem emanates from the fact that $t^{(s)}$ and $t^{(f)}$ can be significantly different from each other. Here we implicitly assume that the compound function $\Delta_k \left(t^{(s)}, t^{(f)} \right)$ is effectively expressed as a linear sum of two different functions with two separate time variables. This is done for mathematical simplicity. Although the function $\Delta_k \left(t^{(s)}, t^{(f)} \right)$ is expressed as a linear sum, the two functions themselves are highly nonlinear with respect to their arguments. Without any loss of generality, we express these two different time scales as a function of a single service time-scale t_s with different coefficients. Using this approach we have

$$\begin{aligned}
\Delta_k \left(t_s \right) &= \Delta_k^{(s)} \left(t_s \right) + \Delta_k^{(f)} \left(t_s \right) \\
&= \underbrace{0.5 e^{-\alpha_k^{(s)} t_s} \frac{(1 + \epsilon_k^{(s)} \cos(\beta_k^{(s)} t_s))}{(1 + \epsilon_k^{(s)})}}_{\Delta_k^{(s)}(t^{(s)})} + \underbrace{0.5 e^{-\alpha_k^{(f)} t_s} \frac{(1 + \epsilon_k^{(f)} \cos(\beta_k^{(f)} t_s))}{(1 + \epsilon_k^{(f)})}}_{\Delta_k^{(f)}(t^{(f)})}.
\end{aligned} \tag{8.5}$$

In Equation (8.5), we assume that the stiffness degradation results from two different processes—one relatively slow and one relatively fast. In the context of aircraft, relatively fast degradation can be manifested in areas which are subjected to repeated loading/unloading such as the landing gears and bulkheads in the fuselage. On the other hand, slow stiffness degradation typically emanates from environmental effects and/or maintenance issues such as corrosion and surface delamination. Numerical values considered for the stiffness degradation are: $\alpha_k^{(s)} = 0.4 \times 10^{-3}$, $\epsilon_k^{(s)} = 0.005$, $\beta_k^{(s)} = 7 \times 10^{-2}$, $\alpha_k^{(f)} = 0.8 \times 10^{-3}$, $\epsilon_k^{(f)} = 0.01$ and $\beta_k^{(f)} = 2 \times 10^{-1}$. Similarly, assume

$$m(t_s) = m_0 (1 + \Delta_m (t_s)), \tag{8.6}$$

where

$$\Delta_m(t_s) = \Delta_m^{(s)}(t_s) + \Delta_m^{(f)}(t_s). \tag{8.7}$$

Analogous to the stiffness degradation case, the mass degradation is also a function of two time-scales—the relatively slower time-scale $\Delta_m^{(s)}(t_s)$ and the relatively faster time-scale $\Delta_m^{(f)}(t_s)$. We have assumed,

$$\Delta_m^{(f)}(t_s) = \epsilon_m \, \text{SawTooth}(\beta_m(t_s - \pi/\beta_m)), \tag{8.8}$$

where $\beta_m = 0.15$ and $\epsilon_m = 0.25$. The slower time-scale is represented as

$$\Delta_m^{(s)}(t_s) = \begin{cases} m_1 & \text{if } t_1 \le t_s < t_2 \\ m_2 & \text{if } t_2 \le t_s < t_3 \\ m_3 & \text{if } t_3 \le t_s < t_4 \\ 0 & \text{elsewhere} \end{cases}. \tag{8.9}$$

From a physical point-of-view, Equation (8.8) can be linked with fuel loading and unloading of an aircraft. On the other hand, Equation (8.9) can be linked with the case where the aircraft drops a bomb during its flight. Schematically, the mass and stiffness degradation are shown in Figure 8.1. For the mass degradation function, the difference in the scale arises from the functional form and the time-discretization necessary for discretizing the process. For the stiffness degradation function, the difference in scale is due to periodicity of the functions. The damping is considered to be constant. The key consideration is that a digital twin of the dynamical system should track changes occurring at multiple-scales by exploiting sensor data measured on the system. Moreover, a digital twin should predict future degradation.

8.2 DIGITAL TWIN FOR MULTI-TIMESCALE DYNAMICAL SYSTEMS

In this section, the proposed digital twin framework for multi-timescale dynamical systems is adumbrated. A schematic representation of the framework is shown in Figure 8.2. The framework proposed has two primary components

1. Data processing by using the physics of the problem (physics-based nominal model)
2. Learning the time-evolution of system parameters by using ML.

Once the material degradation is known, the future responses can be predicted by combining the ML predicted material properties with the problem physics defined by the governing differential equation. To track the multi-timescale nature of the degradation functions, we plan to use the concept of mixture of experts (MoE) where each expert tracks a single time-scale. Based on the success of the Gaussian process (GP) in solving problems having single time-scale in Chapter 7, we propose to use GP as the expert within the MoE framework. The overall framework is named the mixture of experts using Gaussian process (ME-GP). We first outline the details on data processing and then proceed to develop the concept of the proposed ME-GP.

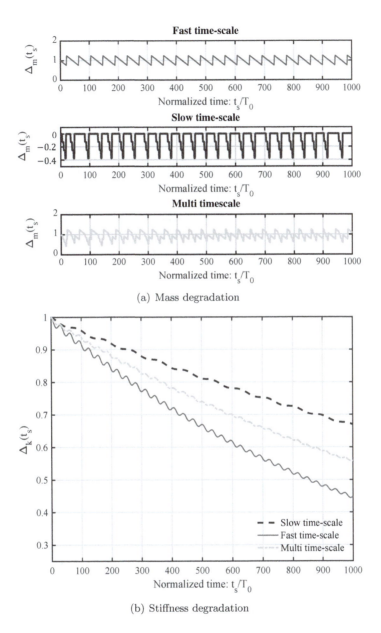

(a) Mass degradation

(b) Stiffness degradation

Figure 8.1 Multi-timescale mass and stiffness degradation functions. The multi-timescale degradation functions are obtained by combining the fast and the slow time-scales shown in each figure. T_0 is the natural time-period of the underlying un-damped system.

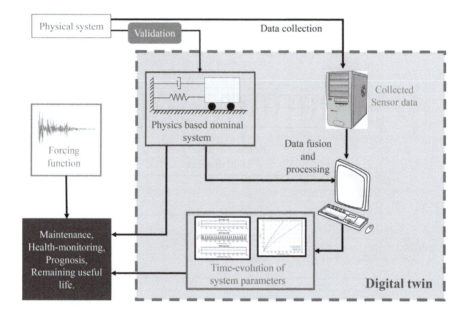

Figure 8.2 Schematic representation of the digital twin. It has three primary building blocks, namely data fusion and processing, determining time evolution of the system parameters and making predictions using the digital twin. This digital twin can be used for several tasks including prognosis, health-monitoring, maintenance, and remaining useful life prediction.

8.2.1 DATA COLLECTION AND PROCESSING

One major enabler in the development of the digital twin technology is the IoT. Advances in IoT have gifted us with several new data collection technologies that, in turn, accelerate the development of the digital twin technology and enables connectivity between and physical and the digital twins. The overall idea of the digital twin technology is buttressed by this connectivity. The connectivity is established by putting sensors on the physical twin to collect data which is then communicated to the digital twin by using cloud computing technology. With advances in the sensor technologies, there are different sensors for collecting different type of responses. In this chapter, we continue to work with the natural frequency of the system. The advantage of using natural frequency lies in the fact that it is a scalar quantity and hence, we can avoid working with a big data-set. We assume that the frequency of the system can be measured in an online setting. Available literature discussed in Chapter 5 illustrates that online frequency measurement is feasible.

In this chapter, three different cases are considered.

1. In the first case, it is assumed that only the stiffness degrades.
2. In the second case, it is assumed the stiffness to be constant; the variability in the observations is due to the variation in the mass.
3. In the third case, it is assumed that both mass and stiffness vary. The collected data needs to be processed differently for each of these three cases.

Details on how the data is processed for each of these three cases is furnished below.

8.2.1.1 Stiffness Degradation

Let us assume that the mass and damping of the nominal model in Equation (8.2) are unchanged and only the stiffness degrades. Accordingly, the equation of motion for this case is expressed as

$$m_0 \frac{\mathrm{d}^2 u(t)}{\mathrm{d}t^2} + c_0 \frac{\mathrm{d}u(t)}{\mathrm{d}t} + k_0 \left(1 + \Delta_k(t_s)\right) u(t) = f(t). \tag{8.10}$$

where all the terms have similar notations as defined before. Note that Equation (8.10) is a special case of Equation (8.2) where t_s is fixed. Solving the characteristic equation, the damped natural frequency of the system can be represented as

$$\lambda_{s_{1,2}}(t_s) = -\zeta_0 \omega_0 \pm i\omega_0 \sqrt{1 + \Delta_k(t_s) - \zeta_0^2}, \tag{8.11}$$

where ω_0 and ζ_0 are respectively the natural frequency and damping ratio of the system at $t_s = 0$. Equation (8.11) can be rearranged as

$$\lambda_{s_{1,2}}(t_s) = -\underbrace{\frac{\zeta_0}{\sqrt{1 + \Delta_{\hat{k}}(t_s)}}}_{\zeta_s(t_s)} \underbrace{\omega_0 \sqrt{1 + \Delta_{\hat{k}}(t_s)}}_{\omega_s(t_s)} \tag{8.12}$$

$$\pm i \underbrace{\omega_0 \sqrt{1 + \Delta_{\hat{k}}(t_s)} \sqrt{1 - \left(\frac{\zeta_0}{\sqrt{1 + \Delta_{\hat{k}}(t_s)}}\right)^2}}_{\omega_{d_s}(t_s)},$$

where $\omega_s(t_s) = \omega_0 \sqrt{1 + \Delta_{\hat{k}}(t_s)}$ i, $\zeta(t_s) = \zeta_0 / \sqrt{1 + \Delta_{\hat{k}}(t_s)}$, and $\omega_{d_s}(t_s) = \omega_s(t_s) \sqrt{1 - \zeta_s^2(t_s)}$ represent the evolution of the natural frequency, damping ratio, and damped natural frequency with t_s, respectively. Since the natural frequency extraction techniques in literature generally extract the damped natural frequency, we have considered it to be the data available from the physical twin. It can be shown [77]

$$\Delta_{\hat{k}}(t_s) = -\tilde{d}_1(t_s) \left(2\sqrt{1 - \zeta_0^2} - \tilde{d}_1(t_s)\right), \tag{8.13}$$

where

$$\tilde{d}_1 (t_s) = \frac{d_1 (\omega_{d_0}, \omega_{d_s} (t_s))}{\omega_0}. \tag{8.14}$$

The function $d_1 (\omega_{d_0}, \omega_{d_s} (t_s))$ in Equation (8.14) is the distance between ω_{d_0} and $\omega_{d_s} (t_s)$

$$d_1 (\omega_{d_0}, \omega_{d_s} (t_s)) = \|\omega_{d_0} - \omega_{d_s} (t_s)\|_2. \tag{8.15}$$

Since the initial damped frequency of the system, ω_{d_0} is known and sensor measurements for $\omega_{d_s} (t_s)$ are available, one can easily compute $\tilde{d}_1 (t_s)$ and $\Delta_{\hat{k}} (t_s)$ at t_s by using Equation (8.14) and substituting it into Equation (8.13). Note that the sensor measurements $\tilde{d}_1 (t_s)$ are likely to be polluted by noise and hence, the estimates for $\Delta_{\hat{k}}$ are also noisy. In this chapter, these noisy estimates, $\Delta_{\hat{k}}$ at discrete time t_s are used for developing the digital twin for the multi-timescale dynamical system.

8.2.1.2 Mass Evolution

In this case, consider that the stiffness and damping of the nominal model in Equation (8.2) are constant. Furthermore, the variation in the observed natural frequency is caused by variation in the mass during the service life. Accordingly, the equation of motion of the physical system becomes

$$m_0 (1 + \Delta_m(t_s)) \frac{d^2 u(t)}{dt^2} + c_0 \frac{du(t)}{dt} + k_0 u(t) = f(t). \tag{8.16}$$

Again, Equation (8.16) is a special case Equation (8.2) where only m varies and the stiffness is constant. Solving for the damped natural eigenfrequencies as before

$$\lambda_{s_{1,2}} (t_s) = -\omega_s (t_s) \zeta_s (t_s) \pm i\omega_{d_s} (t_s), \tag{8.17}$$

where

$$\omega_s (t_s) = \frac{\omega_0}{\sqrt{1 + \Delta_{\hat{m}} (t_s)}}, \tag{8.18a}$$

$$\zeta_s (t_s) = \frac{\zeta_0}{\sqrt{1 + \Delta_{\hat{m}} (t_s)}} \quad \text{and} \tag{8.18b}$$

$$\omega_{d_s} (t_s) = \omega_s (t_s) \sqrt{1 - \zeta_s^2 (t_s)}. \tag{8.18c}$$

are the evolution of natural frequency, damping ratio, and damped natural frequency of the digital twin, respectively. Similar to the stiffness degradation case, we have

$$\Delta_{\hat{m}} (t_s) = \frac{-2\tilde{d}_2 (t_s)^2 + 4\tilde{d}_2 (t_s) \sqrt{1 - \zeta_0^2} - 1 + 2\zeta_0^2}{2 \left(-\tilde{d}_2 (t_s) + \sqrt{1 - \zeta_0^2} \right)^2}$$

$$+ \frac{\sqrt{1 - 4\tilde{d}_2 (t_s)^2 \zeta_0^2 + 8\tilde{d}_2 (t_s) \sqrt{1 - \zeta_0^2}\zeta_0^2 - 4\zeta_0^2 + 4\zeta_0^4}}{2 \left(-\tilde{d}_2 (t_s) + \sqrt{1 - \zeta_0^2} \right)^2} \tag{8.19}$$

where $\tilde{d}_2\left(t_s\right)$ is the equivalent of $\tilde{d}_1$ for the stiffness evolution case. Note that the sensor-based estimates of the damped natural frequencies are noisy. Therefore, the estimated $\Delta_{\hat{m}}\left(t_s\right)$ are also noisy. In this case, we utilize the noisy $\Delta_{\hat{m}}\left(t_s\right)$ at discrete time t_s to construct the digital twin for multi-timescale systems.

8.2.1.3   Mass and Stiffness Evolution

In this case, consider the evolution of mass and degradation of stiffness, simultaneously. The equation of motion in this case is expressed as

$$m_0\left(1+\Delta_m(t_s)\right)\frac{\mathrm{d}^2u(t)}{\mathrm{d}t^2}+c_0\frac{\mathrm{d}u(t)}{\mathrm{d}t}+k_0\left(1+\Delta_k(t_s)\right)u(t)=f(t). \qquad (8.20)$$

All the notations in Equation (8.20) have identical meaning as before. The damped natural eigenfrequencies of this system can are given as

$$\lambda_{s_{1,2}}=-\omega_s\left(t_s\right)\zeta_s\left(t_s\right)\pm i\omega_{d_s}\left(t_s\right), \qquad (8.21)$$

where

$$\omega_s\left(t_s\right)=\omega_0\frac{\sqrt{1+\Delta_{\hat{k}}\left(t_s\right)}}{\sqrt{1+\Delta_{\hat{m}}\left(t_s\right)}} \qquad (8.22a)$$

$$\zeta_s\left(t_s\right)=\frac{\zeta_0}{\sqrt{1+\Delta_{\hat{m}}\left(t_s\right)}\sqrt{1+\Delta_{\hat{k}}\left(t_s\right)}}\quad\text{and} \qquad (8.22b)$$

$$\omega_{d_s}\left(t_s\right)=\omega_s\left(t_s\right)\sqrt{1-\zeta_{ss}^2}. \qquad (8.22c)$$

Here ω_s, ζ_s, and ω_{d_s}, respectively, represent the evolution of natural frequency, damping ratio, and damped natural frequency. Unlike the previous two cases, both $\Delta_m(t_s)$ and $\Delta_k(t_s)$ are unknowns in this case. Therefore, we need two equations to solve for these unknowns. To that end, consider the real and imaginary parts of Equation (8.21) separately to derive the two equations necessary for estimating $\Delta_m(t_s)$ and $\Delta_k(t_s)$. With this setup, the following expression [77] is obtained

$$\Delta_{\hat{m}}\left(t_s\right)=-\frac{\tilde{d}_{\mathcal{R}}\left(t_s\right)}{\zeta_0+\tilde{d}_{\mathcal{R}}\left(t_s\right)}, \qquad (8.23a)$$

$$\Delta_{\hat{k}}\left(t_s\right)=\frac{\zeta_0\tilde{d}_{\mathcal{R}}^2\left(t_s\right)-\left(1-2\zeta_0^2\right)\tilde{d}_{\mathcal{I}}\left(t_s\right)+\zeta_0^2\tilde{d}_{\mathcal{I}}^2\left(t_s\right)}{\zeta_0+\tilde{d}_{\mathcal{R}}\left(t_s\right)}, \qquad (8.23b)$$

where $\tilde{d}_{\mathcal{R}}\left(t_s\right)$ and $\tilde{d}_{\mathcal{I}}\left(t_s\right)$, as before, are distance measures

$$\tilde{d}_{\mathcal{R}}\left(t_s\right)=\frac{d_{\mathcal{R}}\left(t_s\right)}{1+\Delta_{\hat{m}}\left(t_s\right)}, \qquad (8.24)$$

$$\tilde{d}_{\mathcal{I}}\left(t_s\right)=\sqrt{1-\zeta_0^2}-\frac{\sqrt{\left(1+\Delta_{\hat{k}}\left(t_s\right)\right)\left(1+\Delta_{\hat{m}}\left(t_s\right)\right)-\zeta_0^2}}{1+\Delta_{\hat{m}}\left(t_s\right)}.$$

Note that $\Delta_{\hat{m}}(t_s)$ and $\Delta_{\hat{k}}(t_s)$ are estimated from noisy observations of λ and ergo, are noisy. In this case also, we utilize $\Delta_{\hat{m}}(t_s)$ and $\Delta_{\hat{k}}(t_s)$ obtained at discrete time t_s within the digital twin framework for a multi-timescale dynamical system.

8.2.2 MIXTURE OF EXPERTS WITH GAUSSIAN PROCESS

In the next part of the digital twin framework, the processed data is exploited to learn the evolution of the system parameters. Note that the evolution of the system parameters are of multi-timescale nature and hence, is onerous to learn. In this chapter, we propose the use of mixture of experts (MoE) within the digital twin framework. MoE is used to learn the evolution of the system parameters. We state that since each of the experts within MoE learns the evolution at a single scale and hence, MoE can predict the evolution of the system parameters. As experts within the MoE framework, we propose the use of GP. The effectiveness of GP in predicting parameter evolution at a single scale has already been established in Chapter 7.

Suppose, there exist a sequence of observations given as $\boldsymbol{y}_{t_s} \in \mathbb{R}^d$ at discrete time t_s, $s = 1, 2, \ldots, \tau$. For the digital twin problem in this study, $\boldsymbol{y}_{t_s}$ can either be $\Delta_k(t_s)$ and/or $\Delta m(t_s)$ at discrete time t_s. These observations are created from some unknown process possessing multiple time-scales. We assume that the observations are created from M hidden states $x_t^{(m)}$, $m = 1, 2, \ldots, M$, also referred to as experts. These hidden states are typically assumed to be independent and can evolve independent of each other. We assume that the independent hidden states evolve according to a GP

$$x_m | t \sim \mathcal{GP}\left(\mu_m(t; \boldsymbol{h}), \kappa_m(t_1, t_2; \boldsymbol{l})\right), \quad m = 1, 2, \ldots, M, \qquad (8.25)$$

where

$$\mu_m(t) = \boldsymbol{h}^T \boldsymbol{\phi}(t), \qquad (8.26)$$

represents the mean of the GP. $\boldsymbol{h}$ in Equation (8.25) represents the unknown coefficients and $\boldsymbol{\phi}(t)$ represents the basis function vector. κ_m in Equation (8.25) represents the correlation function with length-scale parameter $\boldsymbol{l}$. $\boldsymbol{h}$ and $\boldsymbol{l}$ are called the hyperparameters of the GP. Equation (8.25) can be considered as the prior, parameterized by the hyperparameters $\boldsymbol{h}$ and $\boldsymbol{l}$ in the space of the hidden state. These hidden states can be coupled in a generative manner to obtain the underlying data

$$y_{t_s} = y(t = t_s) = \sum_{m=1}^{M} z_m(t; \boldsymbol{\theta}^g) x_m(t; \boldsymbol{\theta}^e). \qquad (8.27)$$

$z_m(t)$ is the m–th gating function and defined as

$$z_i(t) = \frac{\pi_i \mathcal{N}\left(t | \mu_i, \lambda_i^{-1}\right)}{\sum_{j=1}^{M} \pi_j \mathcal{N}\left(t | \mu_j, \lambda_j^{-1}\right)}, \quad \sum_{j=1}^{M} \pi_j = 1. \qquad (8.28)$$

$\boldsymbol{\theta}^g = [\mu_j, \lambda_j]_{j=1}^M$ are the hyperparameters of the gating function. $\pi_j,\ j = 1, 2, \ldots, M$ in Equation (8.28) represents the mixing coefficient. x_m in Equation (8.27) is the m-th GP expert. $\boldsymbol{\theta}^e = [\boldsymbol{h}_m, \boldsymbol{l}_m]_{m=1}^M$ is the hyperparameter associated with the expert function.

To use the model defined in Equations (8.25)–(8.28), all the hyperparameters must be estimated based on the training data $\mathcal{D} = [\boldsymbol{y}_{t_s}, t_s]$. One way to accomplish this is by maximizing the data likelihood of the model.

$$p\left(y_{t_s}|t_s, \boldsymbol{\theta}, \boldsymbol{\pi}\right) = \sum_{i=1}^M p\left(i|t_s, \boldsymbol{\theta}^g, \boldsymbol{\pi}\right) p\left(y_{t_s}|t_s, \boldsymbol{\theta}^e\right) \tag{8.29}$$

$p\left(i|t_s, \boldsymbol{\theta}^g, \boldsymbol{\pi}\right)$ in Equation (8.29) is the posterior conditional probability, where t_s is assigned to the partition corresponding to the i–th expert, i.e.,

$$p\left(i|t_s, \boldsymbol{\theta}^g, \boldsymbol{\pi}\right) = z_i(t). \tag{8.30}$$

$p\left(\boldsymbol{y}_{t_s}|t_s, \boldsymbol{\theta}^e\right)$, on the other hand, is the probability distribution of the i–th expert and hence is a GP

$$p\left(y_{t_s}|t_s, \boldsymbol{\theta}^e,\right) = \mathcal{N}\left(\mu_i\left(t_s; \boldsymbol{h}\right), \kappa_i\left(t_{s,1}, t_{s,2}; \boldsymbol{l}\right)\right). \tag{8.31}$$

Substituting Equations (8.28) and (8.31) into Equation (8.29), we obtain

$$p\left(y_{t_s}|t_s, \boldsymbol{\theta}, \boldsymbol{\pi}\right) = \tag{8.32}$$

$$\sum_{i=1}^M \frac{\pi_i \mathcal{N}\left(t|\mu_i, \lambda_i^{-1}\right)}{\sum_{j=1}^M \pi_j \mathcal{N}\left(t|\mu_j, \lambda_j^{-1}\right)} \mathcal{N}\left(y_{t_s}|\mu_i\left(t_s; \boldsymbol{h}\right), \kappa_i\left(t_{s,1}, t_{s,2}; \boldsymbol{l}\right)\right).$$

Note that Equation (8.32) is analytically intractable. Using the training samples $\boldsymbol{y}_{t_s}$ and $\boldsymbol{t}_s,\ s = 1, 2, \ldots, \tau$, the likelihood can be expressed as

$$p\left(\boldsymbol{y}_{t_s}|\boldsymbol{t}_s, \boldsymbol{\theta}, \boldsymbol{\pi}\right) = \tag{8.33}$$

$$\prod_{s=1}^\tau \sum_{i=1}^M \frac{\pi_i \mathcal{N}\left(t_s|\mu_i, \lambda_i^{-1}\right)}{\sum_{j=1}^M \pi_j \mathcal{N}\left(t_s|\mu_j, \lambda_j^{-1}\right)} \mathcal{N}\left(y_{t,s}|\mu_i\left(t_s; \boldsymbol{h}\right), \kappa_i\left(t_{s,1}, t_{s,2}; \boldsymbol{l}\right)\right).$$

One way to calculate the parameters in Equation (8.33) is by using maximum likelihood estimator where we maximize the likelihood in Equation (8.33). However, such a procedure can lead to over-fitting. An alternative to the maximum likelihood estimator is to use a Bayesian approach and calculate the posterior distribution of the hyperparameters. However, as the likelihood is intractable for the problem being considered, such an approach is computationally expensive. In this chapter, we adopt a hybrid approach where some of the parameters are treated in a Bayesian manner while for the other parameters, point estimates are computed. More specifically, within the proposed framework, we calculate point-estimates for the mixing coefficients $\boldsymbol{\pi}$. The hyperparameters corresponding to the gating distribution and the experts are treated in a Bayesian manner.

To estimate the hyperparameters using the proposed hybrid approach, we first invoke Bayes rule to calculate the posterior distribution of the hyperparameters θ and π

$$p\left(\theta, \pi | y_s, t_s\right) = \frac{p\left(\pi, \theta\right) p\left(y_s | t_s, \pi, \pi\right)}{p\left(y_s | t_s\right)}, \tag{8.34}$$

where $p\left(\pi, \theta\right)$ represents prior distribution of the hyperparameters and $p\left(y_s | t_s, \pi, \pi\right)$ is obtained from Equation (8.33). Recall that the objective here is to compute point-estimates for the mixing parameters π. This can be accomplished by maximizing the log-posterior for the mixing coefficients

$$\mathcal{L}\left(\pi\right) = \log p\left(\pi | y_{t_s}, t_s\right) = \log \int p\left(\pi, \theta | y_{t_s}, t_s\right) d\theta. \tag{8.35}$$

Unfortunately, this is not simple since Equation (8.35) involves integration over the unknown θ. In this chapter, we propose to use expectation maximization to compute the mixing coefficients by maximizing the log-posterior in Equation (8.35). In expectation maximization, iterations are performed over a series of increasing lower-bound of $\mathcal{L}\left(\pi\right)$ by using the Jensen's inequality

$$
\begin{aligned}
\mathcal{L}\left(\pi\right) = \log p\left(\pi | y_{t_s}, t_s\right) &= \log \int p\left(\pi, \theta | y_{t_s}, t_s\right) d\theta \\
&= \log \int q\left(\theta\right) \frac{p\left(\pi, \theta | y_{t_s}, t_s\right)}{q\left(\theta\right)} d\theta \\
&\geq \int q\left(\theta\right) \log \frac{p\left(\pi, \theta | y_{t_s}, t_s\right)}{q\left(\theta\right)} d\theta \\
&= F\left(q, \pi\right),
\end{aligned}
\tag{8.36}
$$

where $q\left(\theta\right)$ is an auxiliary distribution. It is clear that the equality in Equation (8.36) holds when $q\left(\theta\right) = p\left(\theta | \pi, y_{t_s}, t_s\right)$. Using expectation maximization, π is calculated by iterating over the E-step (expectation step) and the M-step (maximization step).

E-step: Given an estimate of $\pi = \pi^{(s)}$ in step s, we obtain the lower-bound

$$
\begin{aligned}
F\left(q^{(s)}, \pi\right) &= \int q^{(s)}\left(\theta\right) \log p\left(\pi.\theta | y_{t_s}, t_s\right) d\theta \\
&- \int q^{(s)}\left(\theta\right) \log \int q^{(s)}\left(\theta\right) d\theta.
\end{aligned}
\tag{8.37}
$$

M-step: Maximize $F\left(q^{(s)}, \boldsymbol{\pi}\right)$ to update $\boldsymbol{\pi}$.

$$\boldsymbol{\theta}^{(s+1)} = \arg\max_{\boldsymbol{\theta}} F\left(q^{(s)}, \boldsymbol{\pi}\right)$$
$$= \arg\max_{\boldsymbol{\theta}} \left[\mathbb{E}_{q^{(s)}(\boldsymbol{\theta})}\left(\log p\left(\boldsymbol{\pi}, \boldsymbol{\theta} | \boldsymbol{y}_{t_s}, \boldsymbol{t}_s\right)\right)\right]. \tag{8.38}$$

The second equality in Equation (8.38) holds because the second term of $F\left(q^{(s)}, \boldsymbol{\pi}\right)$ is independent of $\boldsymbol{\pi}$. It is important to realize that the optimal distribution $q^{(s)}(\boldsymbol{\theta}) = p\left(\boldsymbol{\theta} | \boldsymbol{\pi}^{(s)}, \boldsymbol{y}_{t_s}, \boldsymbol{t}_s\right)$ is intractable. We propose to use sequential Monte Carlo (SMC) sampler [132] to generate samples from $p\left(\boldsymbol{\theta} | \boldsymbol{\pi}^{(s)}, \boldsymbol{y}_{t_s}, \boldsymbol{t}_s\right)$ so that the expectation in the E-step can be expressed as

$$\mathbb{E}_{q^{(s)}(\boldsymbol{\theta})}\left(\log p\left(\boldsymbol{\pi}, \boldsymbol{\theta} | \boldsymbol{y}_{t_s}, \boldsymbol{t}_s\right)\right) \approx \tag{8.39}$$

$$\sum_{i=1}^{N_s} W^{(s,i)} \log p\left(\boldsymbol{\pi}^{(s)}, \boldsymbol{\theta}^{(s,i)} | \boldsymbol{y}_{t_s}, \boldsymbol{t}_s\right),$$

where $\boldsymbol{\theta}^{(s,i)}$ is the ith sample generated from $p\left(\boldsymbol{\theta} | \boldsymbol{\pi}^{(s)}, \boldsymbol{y}_{t_s}, \boldsymbol{t}_s\right)$, and $W^{(s,i)}$ is the corresponding weight.

Posterior distributions are typically multi-modal and conventional Markov Chain Monte Carlo (MCMC) [134] may get trapped in a local mode. This results in long mixing time which makes the process inefficient. One algorithm that addresses this issue is the SMC sampler [63, 132]. SMC provides a parallelizable framework which efficiently extracts samples from multi-modal posterior distributions. The concept of annealing is introduced to create auxiliary distributions. We peregrinate from the prior to the posterior through these auxiliary distributions; this ensures a smooth transition from the tractable prior to the intractable posterior. It can be shown that samples extracted using SMC converge asymptotically to the target distribution [132].

To apply SMC to approximate the E-step of the expectation maximization algorithm, we first express $p\left(\boldsymbol{\theta} | \boldsymbol{\pi}^{(s)}, \boldsymbol{y}_{t_s}, \boldsymbol{t}_s\right)$ as

$$p\left(\boldsymbol{\theta} | \boldsymbol{\pi}^{(s)}, \boldsymbol{y}_{t_s}, \boldsymbol{t}_s\right) \propto p\left(\boldsymbol{\theta}\right) p\left(\boldsymbol{y}_{t_s} | \boldsymbol{t}_s.\boldsymbol{\pi}^{(s)}, \boldsymbol{\theta}\right). \tag{8.40}$$

where $p\left(\boldsymbol{\theta}\right)$ is the prior and $p\left(\boldsymbol{y}_{t_s} | \boldsymbol{t}_s.\boldsymbol{\pi}^{(s)}, \boldsymbol{\theta}\right)$ is the likelihood of the model defined in Equation (8.33). In this chapter, we set the prior as a multivariate Gaussian distribution with zero mean and identity covariance matrix. Therefore, the parameters $\boldsymbol{\theta}$ are independent in the prior. For ease of representation, we express the likelihood in a compressed form as $p\left(\mathcal{D} | \boldsymbol{\theta}\right)$ and the posterior of $\boldsymbol{\theta}$ as $p_n(\boldsymbol{\theta})$. With this nomenclature, Equation (8.40) is written as

$$p_n\left(\boldsymbol{\theta}\right) \propto p\left(\boldsymbol{\theta}\right) p\left(\mathcal{D} | \boldsymbol{\theta}\right) \tag{8.41}$$

Based on Equation (8.41), following auxiliary distribution in SMC is formulated

$$p_t\left(\boldsymbol{\theta}\right) \propto p\left(\boldsymbol{\theta}\right) p^{\gamma_t}\left(\mathcal{D} | \boldsymbol{\theta}\right), \tag{8.42}$$

where $t = 0, 1, \ldots, n$ and $0 = \gamma_0 < \gamma_1 < \cdots < \gamma_n = 1$ are the annealing parameters. Using the SMC sampler, samples are extracted from such a sequence of probability distribution by utilizing importance sampling and re-sampling. At step t, the idea is to generate a sufficient collection of $\left\{ \boldsymbol{\theta}_r^{(i)}, \boldsymbol{w}_r^{(i)} \right\}$, $i = 1, \ldots, N_s$ so that the empirical distribution converges asymptotically to the target distribution $p_r(\boldsymbol{\theta})$. Sampling at $t = 0$ is trivial (as we sample the prior). From $t = 1$ onward, we apply importance sampling sequentially to the auxiliary distributions. A predefined Markov transition kernel is used. Assuming, at step $t - 1$, N_s samples $\left\{ \boldsymbol{\theta}_{t-1}^{(i)} \right\}$, $i = 1, \ldots, N_s$ are created according to the proposal distribution φ_{t-1}, a kernel K_t with invariant distribution p_t is proposed such that the new samples are marginally distributed as [190]

$$\varphi_t = \int \varphi_{t-1} K_t (\boldsymbol{\theta}, \boldsymbol{\theta}') \, d\boldsymbol{\theta}. \tag{8.43}$$

Following [132], we utilize the Metropolis-Hasting kernel with invariant distribution p_t to move the samples based on a random walk proposal

$$\varphi_t = \mathcal{N} \left(\boldsymbol{\theta}_{r-1}^{(i)}, \mathbf{v}^{(i)} \right), \tag{8.44}$$

where $\mathbf{v}^{(i)}$ is the covariance matrix. To represent the discrepancy between the proposal distribution φ_t and the target distribution p_t at step t, $0 < t \leq n$, unnormalized importance weights $w_t^{(i)}$ are created.

$$w_t^{(i)} = w_t^{(i-1)} \frac{p_t \left(\boldsymbol{\theta}_{t-1}^i \right)}{p_{t-1} \left(\boldsymbol{\theta}_{t-1}^i \right)} \tag{8.45}$$

The calculated weights are normalized as

$$W_t^{(i)} = \frac{w_t^{(i)}}{\sum_{j=1}^{N_s} w_t^{(j)}} \tag{8.46}$$

As mentioned in [121, 132], the SMC sampler degenerates and the variance of the importance weight increases. In this chapter, we measure the degeneracy based on the effective sample size (ESS) [190]

$$\mathrm{ESS}_t = \left(\sum_{t=1}^{N_s} \left(W_t^{(i)} \right)^2 \right)^{-1}. \tag{8.47}$$

We consider degeneracy to have taken place if

$$\mathrm{ESS}_t < \mathrm{ESS}_{\min}, \tag{8.48}$$

where $\mathrm{ESS}_{\min}$ represents the threshold. In this chapter, we have defined $\mathrm{ESS}_{\min} = c \times N_s \, (c < 1)$. In case, $\mathrm{ESS}_t < \mathrm{ESS}_{\min}$, resampling is carried out to ameliorate the degeneracy of the sampler. Once samples corresponding to the target distribution are obtained, we utilize them to calculate the expectation in the E-step of the expectation maximization algorithm. The steps involved in the SMC sampler are adumbrated in Algorithm 3. The steps in-

---

Algorithm 3 Sequential Monte Carlo sampler

Input: Number of samples to generate N_s, the prior distribution $p(\boldsymbol{\theta})$, the number of steps n and the threshold parameters c

Initialize N_s particles $\boldsymbol{\theta}_0^{(i)}$, $i = 1, \ldots, N_s$ by directly sampling the prior distribution $p(\boldsymbol{\theta})$ and set the corresponds weights to be one, $w_0^{(i)} = 1$.

for $t = 1, \ldots, n$ **do**

    for $i = 1, \ldots, N_s$ **do**
    Sample $\boldsymbol{u}_i$ from uniform distribtuion $\mathcal{U}(\mathbf{0}, \mathbf{I})$.

    Sample $\tilde{\boldsymbol{\theta}}$ from the proposal distribution $\mathcal{N}\left(\boldsymbol{\theta}_{t-1}^{(i)}, \mathbf{v}_i\right)$.

$$u_i < \min\left\{\frac{p_t(\tilde{\boldsymbol{\theta}})}{p_t\left(\boldsymbol{\theta}_{t-1}^{(i)}\right)}\right\} \quad \boldsymbol{\theta}_t^{(i)} \leftarrow \tilde{\boldsymbol{\theta}} \quad \boldsymbol{\theta}_t^{(i)} \leftarrow \boldsymbol{\theta}_{t-1}^{(i)}$$

    end for
Set weights of each particle according to Equations (8.45) and (8.46).

    Compute ESS_t using Equation 8.47.

    Resample if $\text{ESS}_t < \text{ESS}_{\min}$.
end for
Use $\boldsymbol{\theta}_n^{(i)}$ and $W_n^{(i)}$, $i = 1, \ldots, N_s$ to compute the expectation in the E-step of the expectation maximization algorithm.

---

volved in training the proposed MoE using GP algorithm are adumbrated in Algorithm 4

Once the hyperparameters of the proposed model are established by using the proposed approach, we proceed to make predictions using the proposed approach. Since we consider a partially Bayesian approach to obtain the hyperparameters $\boldsymbol{\theta}$, it is possible to use the same approach to make *probabilistic* predictions. Suppose, we are interested in obtaining y_{t^*} at time-step t^*. This can be obtained by calculating the posterior predictive distribution.

$$p(y_{t^*}|t^*, \mathcal{D}, \pi^*) = \int_\theta \sum_{i=1}^{M} \underbrace{p(i|t^*, \pi_i^*, \theta_i^g)}_{\text{gating}} \underbrace{p(y^*|t^*, \theta_i^e)}_{\text{expert}} \underbrace{p(\theta_i^e, \theta_i^g|\mathcal{D})}_{\text{posterior}} \, d\boldsymbol{\theta}. \quad (8.49)$$

Algorithm 4 Mixture of experts using Gaussian process

Input: Number of experts M, the training data $\mathcal{D} = [\boldsymbol{y}_{t_s}, \boldsymbol{t}_s]$, $s = 1, \ldots, \tau$, initial values of mixing coefficients $\boldsymbol{\pi}^{(i)}$ and threshold ϵ.

$\boldsymbol{\pi} \leftarrow \boldsymbol{\pi}^{(i)}$.

$\lambda = 10\epsilon$.

repeat$\lambda \leq \epsilon$
$\boldsymbol{\pi}_s \leftarrow \boldsymbol{\pi}$.

Compute $F\left(q^{(s)}, \boldsymbol{\pi}\right)$ using SMC sampler (Algorithm 3).

Update $\boldsymbol{\pi}$ by solving the optimization problem in Equation 8.38.

Compute error threshold
$$\lambda = ||\boldsymbol{\pi} - \boldsymbol{\pi}_s||_2$$

Outcome: Optimized $\boldsymbol{\pi}$, N_s samples and corresponding weights from the posterior of $\boldsymbol{\theta}$ $\boldsymbol{\theta}_n^{(i)}$, $W_n^{(i)}$, $i = 1, \ldots, N_s$.

The integral above can be approximated by using Monte Carlo integration. In particular, we utilize the samples extracted from the posterior along with the correspond weights and the EM estimate of $\boldsymbol{\pi}^*$ to draw samples from the posterior predictive distribution in Equation (8.49).

8.2.3 ALGORITHM

We now discuss how the components discussed in Equations (8.2.1) and (8.2.2) interact with one another within the digital twin framework shown in Figure 8.2, and how the digital twin enhanced with ME-GP can be used for multi-timescale dynamical systems. Given a physical system, the first step toward constructing a digital twin is to develop a physics-driven nominal model for the system. For the current chapter, the nominal model is represented by Equation (8.2). Next, the collected responses (damped natural frequencies of the system) are processed by using the procedure outlined in Section 8.2.1. To be more specific, we process the collected damped natural frequencies to obtain change in the mass, $(\Delta_m(t_s))$ and stiffness $(\Delta_k(t_s))$ of the system. In the third step, the time-evolution of mass and stiffness, $\eta : t \to \Delta_k, \Delta_m$ is learnt by using ME-GP. Finally, using the trained ME-GP, we calculate the future mass and stiffness, substitute them into the nominal model and solve it to get the future responses of interest. These future responses of interest are useful for health-monitoring, computing remaining useful life, devising a maintenance strategy, and identifying defects and/or cracks in the system. The details on the proposed algorithm are provided in Algorithm 5.

Algorithm 5 Proposed digital twin

Input: Nominal model and damped natural frequency of the physical system at different time-instants, $\mathcal{D} = [\lambda_s, t_s]$, $s = 1, \ldots, \tau$.

Process the collected data to obtain $\Delta_k(t_s)$ and/or $\Delta_m(t_s)$ at t_s (See Section 8.2.1).

Use ME-GP to learn the time-evolution of Δ_m and/or Δ_k (See Section 8.2.2).

Obtain $\Delta_k(t^*)$ and/or $\Delta_m(t^*)$ at t^*, $t^* > \tau$ (See Section 8.2.2).

Substitute $k^* = (1 + \Delta_k(t^*))$ and/or $m^* = (1 + \Delta_m(t^*))$ into the nominal model and solve it to obtain responses expected in the future.

Take engineering decision.

Repeat steps (2)–(6) as more data becomes available

The proposed digital twin framework in this chapter has numerous advantages.

1. The framework proposed uses both physics-driven model (ordinary and partial differential equations) and data-driven models (ME-GP). The physics-driven model ensures extrapolatibility of the proposed digital twin. On the other hand, the data-driven model ensures that the proposed digital twin is not limited by the facts that there may be missing physics.
2. Including the physics-based model also enables us to predict other responses of interest. For instance, although we only have sensor information about the damped natural frequency of the system, the proposed digital twin can easily predict other responses such as strains, displacements, and velocity.
3. The fact that we utilize ME-GP enables the digital twin to track even multi-timescale dynamical systems such as the one considered in this chapter.

8.3   ILLUSTRATION OF THE PROPOSED FRAMEWORK

In this section, the performance, utility, and applicability of the proposed digital twin framework for multi-timescale dynamical systems is presented. More specifically, we present results for the problem defined in Section 8.1. Three cases as defined in Section 8.2 are considered. As already stated, it is assumed that we can obtain the damped natural frequency of the system at different (slow) time-steps. The objective is to learn the time-evolution of the mass and/or the stiffness. Once the time-evolution of the mass and/or stiffness is established, the same can be used to determine the response of interest at a given time-instant by solving the physics-driven nominal model. We illustrate how the proposed digital twin can be applied to learn the parameters (mass and stiffness) in the past (interpolation) as well as in the future (extrapolation). Lastly, to illustrate the onerousness of a multi-timescale dynamical system and to showcase the necessity of the ME-GP, the results obtained using the proposed digital twin are compared with those obtained using a GP based digital twin [30].

8.3.1 DIGITAL TWIN VIA STIFFNESS EVOLUTION

First, consider the case where the variation in the measured natural frequency is caused by the degradation in the stiffness of the system. The evolution of stiffness is assumed to be multi-timescale in nature as shown in Figure 8.1(b) (the multi-timescale one). However, neither the pattern of the time-evolution nor the number of scales present in the data is available a-priori. The sensor data is assumed to be transmitted intermittently at a certain regular time-interval. To simulate a realistic scenario, the damped natural frequency data is corrupted with white Gaussian noise having a standard deviation σ_0. The frequency of data availability depends on several factors including the bandwidth of the transmission system and cost of data collection. Therefore, users of the digital twin are not only interested in the behavior of the system at a future time, but also in the behavior at an intermediate time.

Consider the availability of N_s observations of the damped natural frequency $\lambda_s(t_s)$ equally spaced in time $t_s \in [0, \tau]$. Within the proposed digital twin framework, $\lambda_s(t_s)$ is processed based on the procedure described in Section 8.2.1.1 to extract the change in the stiffness $\Delta_{\tilde{k}}(t_s)$. Then, using t_s as the input data and $\Delta_{\tilde{k}}(t_s)$ as the output data, the ME-GP model is trained by using SMC and expectation maximization by using Algorithms 3 and 4. Now the number of scales present in the data typically unknown for a real-life problem. Four experts (different from the two scales that are actually present) are used within the ME-GP framework. Automatic relevance determination based Matern covariance function and quadratic mean function is considered for all the experts. The threshold parameter C is set as 0.85 [42, 106] and 1000 samples are created by using the SMC sampler. The trained ME-GP model is applied as a surrogate to the unknown degradation process. The stiffness at a given time t^* can be determined by using Equation 8.3 where $\Delta_k(t_s) \approx \Delta_{\tilde{k}}(t^*)$ is computed by using the trained ME-GP model. Substituting the estimated stiffness into the nominal model in Equation 8.1 predicts any response of interest at t^*. Moreover, as more data becomes available, the ME-GP model is updated.

Figure 8.3 shows the variation of Δ_k with normalized time. We have 200 measurements equally spaced throughout the service life of the system. The measured data is clean (i.e., no noise). We observe that for this case, the proposed digital twin and the GP-based digital twin yield identical results. In other words, in the presence of a sufficient amount of (clean) data throughout the service life of the system, the proposed ME-GP-based digital twin transforms into a simple GP-based digital twin. However, in a real-life setting, access to life-time data is rare. Furthermore, the data available is typically corrupted by some form of noise.

Next, more realistic cases are considered. More specifically, we consider cases where we only have measurements in a certain observation time-window, $[0, \tau]$ and we want to predict the evolution of stiffness at a time t^* where $t^* > \tau$. Moreover, the data gathered from the physical system is considered

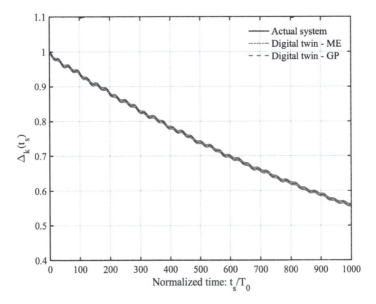

Figure 8.3 Results obtained using GP-based and ME-GP-based digital twin for the case where only stiffness changes. Ideal scenario is considered where clean data throughout the service life is available.

to be noisy. Figure 8.4 shows the performance of digital twin for cases where $\tau = [150, 250, 550]$. For $\tau = 150$, we have 35 measurement data while for the other two cases, 50 sensor measurements are available. For all the three cases, the sensor measurements are corrupted by white Gaussian noise with standard deviation $\sigma_0 = 0.005$. The observed and unobserved regime are also marked in the figure by using a vertical line. Similar to Figure 8.3, results have been determined by using the proposed digital twin and the GP-based digital twin. For $\tau = 150$ (Figure 8.4(a)), the proposed ME-GP-based digital twin yields excellent result up to $t_s/T_0 \approx 600$, which is almost four times the observation time window. Even beyond $t_s/T_0 = 600$, the results obtained from the proposed digital twin are found to be satisfactory. For clarity, the predictive uncertainty for this case is not reported. The GP-based digital twin, on the other hand, yields erroneous results almost immediately after the observation time-window.

A necessary characteristic of digital twin is its ability to update itself as more data becomes available. Figures 8.4(b) and 8.4(c) show the results when more data becomes available and the the digital twin is updated accordingly. For Figure 8.4(b), we consider that we have access to 50 data points equally spaced in the time-window $[0, 250]$. Similarly, in Figure 8.4(c), we have access to 50 data points equally spaced in the time-window $[0, 550]$. We observe that with longer observation window, the digital twin can capture the evolution

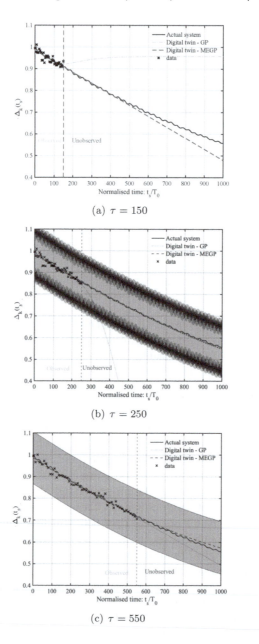

(a) $\tau = 150$

(b) $\tau = 250$

(c) $\tau = 550$

Figure 8.4 Results obtained using GP and ME-GP based digital twin with observations up to $\tau = [150, 250, 550]$ and noise variance $\sigma_0 = 0.005$. For $\tau = 150$, we have 35 sensor data whereas for $\tau = [250, 550]$, we have 50 sensor data (shown using cross). The observed and unobserved regime are differentiated by using a vertical line.

of stiffness up to $t_s/T_0 = 1000$ (which is assumed to be the service life of the system). We also note that the predictive uncertainty for both these cases envelops all the observed data points. It can thus be surmised that the uncertainty in the system caused by limited and noisy measurements is adequately captured. The GP-based digital twin is also updated for the two cases. Although the GP based digital twin yields erroneous result for $\tau = 250$, it is found to give satisfactory results for $\tau = 550$. However, the predicted results of the ME-GP based digital twin are superior.

Lastly, we study the case when the noise variance is high. Figure 8.5 presents the results corresponding to $\sigma_0 = 0.015$. Same three cases as Figure 8.4 are considered. We see that for $\tau = 150$, the prediction of the ME-GP-based digital twin oscillates around the actual solution (Figure 8.5(a)). This oscillatory behavior likely occurs because ME-GP overfits to the noise in the observations. The GP-based digital twin is found to give erroneous results. The digital twin models are then updated by collecting data up to $\tau = 250$ and $\tau = 550$. For both these cases, ME-GP-based digital twin yields reasonably good predictions. However, due to the increase in the noise variance, the predicted results are inferior to those shown in Figure 8.4. This clearly shows the need to collect clean data from the physical system. The predictive uncertainty for $\tau = 250$ and $\tau = 550$ is also shown. We observe that with increase in the observation time-window, the predictive uncertainty at $t/T_0 = 1000$ is almost identical. This occurs because the noise in the observations for both the cases are identical and the last observation in Figure 8.5 (c) is still significantly removed from $t/T_0 = 1000$. Furthermore, note that the predictive uncertainty for Figure 8.5 is lesser as compared to Figure 8.4. Such a counter intuitive result can be ascribed to the fact that with more noise, it becomes more difficult for all ML models including ME-GP to correctly capture the trend. A few observation points in Figure 8.5 lie outside the predictive uncertainty; this indicates that the ME-GP model has become over-confident for this case. The GP-based digital twin for all the three cases in Figure 8.5 are found to proffer erroneous results beyond the observation window. This established the superiority of the proposed ME-GP over GP, particularly when predicting the future.

8.3.2 DIGITAL TWIN VIA MASS EVOLUTION

Next, we consider the case corresponding to Ref. 8.2.1.2 where the variation in the damped natural frequency of the system is because of the variation in its mass. The evolution of mass is found to be of multi-timescale nature as shown in Ref. 8.1(a). However, similar to the stiffness degradation case, neither the pattern of time-evolution nor the number of scales present in the data is available to us a-priori. We again assume that the sensor data are transmitted intermittently at a fixed time interval. To conjure a realistic scenario, the damped natural frequency data is polluted with white Gaussian noise having a standard deviation σ_0. The objective is to apply the proposed digital twin to predict the multi-timescale evolution of mass.

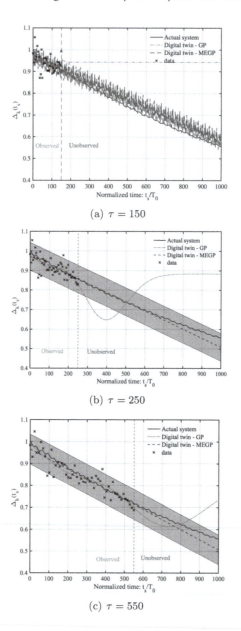

Figure 8.5 Results obtained using GP and ME-GP-based digital twin with observations up to $\tau = [150, 250, 550]$ and noise variance $\sigma_0 = 0.015$. For observation up to 150, we have 35 sensor data whereas for observation up to 250 and 550, we have 50 sensor data (shown using cross). The observed and unobserved regime are differentiated by using a vertical line.

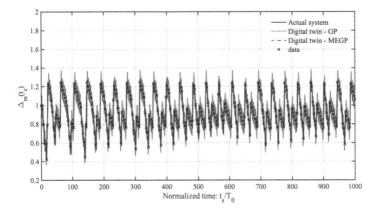

Figure 8.6 Results obtained using GP-based and ME-GP-based digital twin for the case where only mass changes. Ideal scenario is considered where equally spaced clean data throughout the service life is available.

Consider that there exist N_s observations of the damped natural frequency $\lambda_s(t_s)$ equally spaced in time $t_s \in [0, \tau]$. Using the procedure discussed in Section 8.2.1.2, we first process $\lambda_s(t_s)$ to determine the change in the mass $\Delta_{\tilde{m}}(t_s)$. Thereafter, using t_s as the input and $\Delta_{\tilde{m}}(t_s)$ as the output, we train an ME-GP model by using Algorithms 3 and 4. The setup for the algorithms are kept close to that described in Section 8.3.1 and the trained ME-GP model is treated as a surrogate of the unknown Δ_m. For obtaining the response of the system at t^*, we can use the trained ME-GP model to determine the updated m at t^*, substitute it into the nominal model, and then solve it to get the responses of interest. As more data becomes available, the digital twin is updated.

Figure 8.6 shows the variation of Δ_m with normalized time t_s/T_0. These results correspond to the ideal case where we have 300 measurements equally spaced throughout the service life of the system. The data available are also bereft of any noise. We see that the proposed digital twin and the GP-based digital twin yields identical results. In other words, for this case, the proposed ME-GP-based digital twin reduces to the GP based digital twin proposed in Chapter 7. However, note that this is an ideal scenario. In a more realistic setting, the data collected is noisy. Furthermore, data over the complete service-life is seldom available.

Next we address a more realistic setup where we have data over a certain time-window $[0, \tau]$ and the goal is to predict Δ_m at a time t^* where $t^* > \tau$. The data collected is corrupted by white Gaussian noise. Figure 8.7(a) shows the digital twin predicted results corresponding to $\tau = 150$. The data collected is corrupted by white noise with $\sigma_0 = 0.005$. The observed and unobserved regimes are marked by a vertical line. Results using GP-based digital twin are also created for the sake of comparison. For $\tau = 150$, the

ME-GP model trained from only 75 data points provides a reasonable prediction of the evolution of mass over the the service life of the system. More specifically, the primary crest of the response is accurately captured throughout the service life. However, as time increases, the digital twin over-estimates the trough. The proposed approach, being Bayesian in nature, also yields the predictive uncertainty as indicated by the shaded plot. Throughout the service life, the true solution lies within the shaded plot, indicating that the uncertainty emanating from limited and noisy data is properly captured. The GP-based digital twin provides an excellent prediction in $t_s/T_0 = [0, 150]$. However, it yields erroneous prediction almost immediately beyond the observation time-window.

Figure 8.7(b) shows the results predicted using the digital twins for $\tau = 550$. Note that 175 equally spaced observations over the time-window are collected. The data collected is polluted with white Gaussian noise having $\sigma_0 = 0.005$. Within the observation time-window, GP is found to give better results as compared to proposed ME-GP. However, beyond the observation window, GP fails to yield a reasonable prediction. ME-GP, on the other hand, provides reasonable prediction even beyond the observation window.

Lastly, consider the case when the noise variance is larger. Figure 8.8 illustrated the results corresponding to $\sigma_0 = 0.015$. The predicted time-evolution of mass is observed to be similar to that predicted in Figure 8.7. The predictive uncertainty is observed to increase for this case. This can be ascribed to the increase in the noise in the data. GP-based digital twin is unable to properly predict the time-evolution beyond the observation window.

8.3.3 DIGITAL TWIN VIA MASS AND STIFFNESS EVOLUTION

Lastly, we show the performance of the proposed digital twin when both mass and stiffness evolve with time. Evolution of both mass and stiffness is of multi-timescale nature as illustrated in Figure 8.1. Similar to the previous cases, the sensor data is assumed to be transmitted intermittently at a fixed time interval. Considering that we have N_s observations of the damped natural frequency, the digital twin first processes this data to obtain Δ_m and Δ_k. Details on the data processing step are given in Section 8.2.1.3. Thereafter, ME-GP is applied to learn the time evolution of mass and stiffness. The parameters for the ME-GP algorithm are kept identical as that described in Section 8.3.3. To determine response at a given time-instant t^*, we first obtain the mass and stiffness by using the trained ME-GP model as a surrogate. Thereafter, the responses of interest are determined by substituting the ME-GP predicted mass and stiffness into the nominal model and solving it. Similar to the previous two cases, we only proffer the performance of the digital in predicting the time evolution of mass and stiffness. The logic of this argument is that if the time evolution of mass and stiffness is accurately captured, the responses predicted must also be accurate.

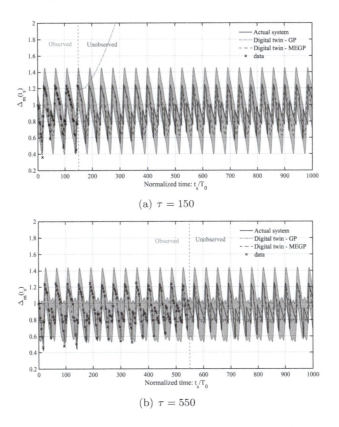

(a) $\tau = 150$

(b) $\tau = 550$

Figure 8.7 Results obtained using GP and ME-GP-based digital twin with observation window of $[0, 150]$ and $[0, 550]$, and $\sigma_0 = 0.005$. For observation window of $[0, 150]$, we have 75 sensor data whereas for observation window of $[0, 550]$, we have 175 sensor data (shown using cross). The observed and unobserved regime are differentiated by using a vertical line.

Since the proposed digital twin yields almost exact result for the ideal scenario for the previous two cases, we directly embark on the realistic case-studies. First, we address the case where the observation window is $[0, 150]$. Within this time-window, it is assumed that we get access to 75 equally spaced sensor measurements. The data collected is polluted by a white Gaussian noise with $\sigma_0 = 0.025$. Figure 8.9 shows the evolution of Δ_m and Δ_k estimated using the digital twin. For Δ_m, it is found that the digital twin gives reasonable prediction throughout the service life of the system. However, for Δ_k, the results are seen to deviate beyond $t_s/T_0 = 350$. Since the proposed approach is Bayesian, the predictive uncertainty has also been determined. For Δ_m, the predictive uncertainty envelops the true behavior throughout the service life; this indicates that the uncertainty emanating from noise and limited data is

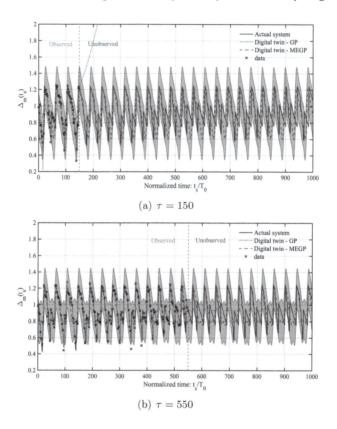

(a) $\tau = 150$

(b) $\tau = 550$

Figure 8.8 Results obtained using GP and ME-GP based digital twin with observation window of $[0, 150]$ and $[0, 550]$, and $\sigma_0 = 0.015$. For observation window of $[0, 150]$, we have 75 sensor data whereas for observation window of $[0, 550]$, we have 175 sensor data (shown using cross). The observed and unobserved regime are differentiated by using a vertical line.

appropriately captured. However, for Δ_k, the correct solution lies outside the envelope beyond $t_s/T_0 = 600$. This clearly shows that the proposed digital twin is over-confident when $t_s/T_0 > 600$. The GP-based digital twin for both Δ_m and Δ_k gives erroneous results beyond $t_s/T_0 = 200$.

The performance of the digital twin is improved by doing an investigation with additional sensor data to the model. More specifically, 120 and 150 equally spaced observations are provided within the same observation window $[0, 150]$. The results are shown in Figure 8.10. As the predictions for Δ_m are already reasonable in Figure 8.11, only the results corresponding to Δ_k are shown. We see that with an increase in the number of observations, the digital twin predictions become nearer to the actual solution. The predictive uncertainty is also found to improve as it envelops the true solution.

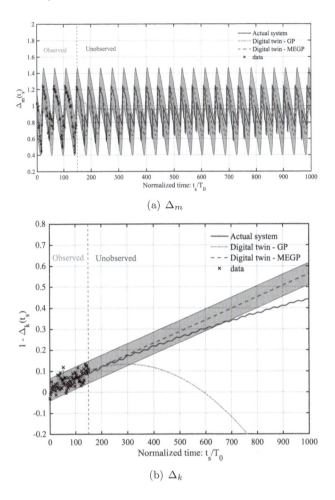

(a) Δ_m

(b) Δ_k

Figure 8.9 Digital twin predicted responses, Δ_m and Δ_k. The digital twin is trained using 75 equally spaced sensor measurements in observation window $[0, 150]$. The sensor data are contaminated by white Gaussian noise with $\sigma_0 = 0.025$. The observed and unobserved regime are shown by a vertical line.

As the final case study, an investigation is carried out by increasing the observation window to $[0, 350]$. However, the number of sensor observations is kept fixed at 75. The result is shown in Figure 8.11 With this setup, the digital twin predicts the time evolution of the stiffness almost perfectly. This illustrates the significance of obtaining data over a longer time-span. For all the cases addressed, the GP-based digital twin fails to proffer accurate prediction beyond the observation window.

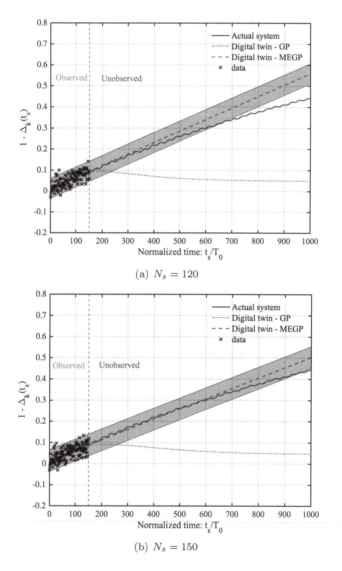

(a) $N_s = 120$

(b) $N_s = 150$

Figure 8.10 Digital twin performance with increase in number of sensor measurements over the observation window $[0, 150]$. The sensor data are contaminated with white Gaussian noise having $\sigma_0 = 0.025$.

8.4 SUMMARY

Digital twins of dynamical systems encountered in engineering and technology should be amenable to the use of multiple time-scales. The solution of such systems is onerous from a computational point-of-view as we need a time-step of the order of the fastest scale. As a result, tasks such as

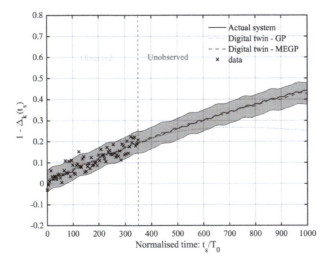

Figure 8.11   Performance of digital twin for observation time-window $[0, 350]$. Note that 75 sensor data are available within the specified time-window. The data are contaminated by white Gaussian noise with $\sigma_0 = 0.025$.

health-monitoring, damage prognosis, and remaining useful-life prediction can become intractable. To address this issue, a ML-based digital twin framework for multi-timescale dynamical systems is presented. The proposed digital twin has two main components: (a) a physics-driven nominal model [typically represented by ordinary or partial differential equation(s)] and (b) a data-driven machine learning model. We use the physics-driven nominal model for data processing and predictions and the ML model for learning the time-evolution of the system parameters. We use a mixture of experts (MoE) as the ML model. As an expert within the MoE framework, we use GP. The basic idea is that expert will monitor the temporal evolution of the system parameters at a single scale. To learn the hyperparameters of the proposed model, an algorithm based on expectation-maximization and sequential Monte Carlo sampler is proposed. The proposed training algorithm is of hybrid type where some of the parameters are addressed in a Bayesian sense while the point-estimates for others are provided.

The results obtained using the proposed approach show its ability to predict the time-evolution of the system parameters, even outside the observation time-window. However, extremely sparse and highly noisy data can deleteriously affect the performance of the proposed framework. Moreover, gathering data over a longer observation window can dramatically improve the performance of the proposed framework. The proposed framework being partially Bayesian quantifies the uncertainty due to limited and noisy data. In most of the cases, the predictive uncertainty envelops the true solution, indicating that the uncertainty is well captured. However, for some cases, the true solution is found to lie outside the envelope. This is probably caused by over-fitting and employing a fully-Bayesian framework can ameliorate this issue.

Despite the several findings in the current chapter, there are three possible extensions that need to be addressed in future.

1. First, the illustration has been conducted using a single SDOF system. While this facilitates the understanding the functionality of the proposed framework, further investigation on MDOF systems is necessary. For MDOF systems, a major challenge lies in the data-processing step. This is because deriving closed-form relation between frequency measurements and system parameters is difficult, if not impossible.
2. Secondly, we have considered the damped natural frequency to be our observation. However, this is a derived quantity based on strain measurements. It is necessary to extend the framework to directly learn the evolution of mass and stiffness from time-history measurements.
3. Thirdly, the framework proposed in this chapter does not utilize the system physics and the data simultaneously; instead the physics and data are utilized in a decoupled way. There are methods rooted in physics-informed ML framework where the physics and data approaches can be fused. Potential of such methods within the digital twin framework needs to be investigated.

9 Digital Twin of Nonlinear MDOF Systems

Most real systems have multiple degrees of freedom. Additionality, real physical systems may be nonlinear. These two issues are addressed in the current chapter thereby taking the digital twin development process to a high level of fidelity. This chapter is adapted from Ref. [79].

9.1 PHYSICS-BASED NOMINAL MODEL

In this section, we present the nominal dynamic system and the digital twin corresponding to this model. The nominal model is the "initial model" of a DT, also called the baseline model in this book. For engineering systems, we can consider a nominal model to be a numerical model of the system when it is manufactured. A DT captures the journey from the nominal model to its updates based on the data collected from the system. In this section, the key ideas for developing DT of nonlinear MDOF systems are adumbrated.

9.1.1 STOCHASTIC NONLINEAR MDOF SYSTEM: THE NOMINAL MODEL

Consider an $N-$DOF stochastic nonlinear system having governing equations given by:

$$\mathbf{M}_0\ddot{\boldsymbol{X}} + \mathbf{C}_0\dot{\boldsymbol{X}} + \mathbf{K}_0\boldsymbol{X} + \boldsymbol{G}\left(\boldsymbol{X},\boldsymbol{\alpha}\right) = \boldsymbol{F} + \boldsymbol{\Sigma}\dot{\boldsymbol{W}} \tag{9.1}$$

where $\mathbf{M}_0 \in \mathbb{R}^{N \times N}$, $\mathbf{C}_0 \in \mathbb{R}^{N \times N}$, and $\mathbf{K}_0 \in \mathbb{R}^{N \times N}$, respectively, represent the mass, damping, and (linear) stiffness matrix of the system. $\boldsymbol{G}\left(\cdot,\cdot\right) \in \mathbb{R}^N$, on the other hand, represents the nonlinearity present in the system. $\boldsymbol{F}$ in Equation (9.1) represents the deterministic force and and $\dot{\boldsymbol{W}}$ (Wiener derivative) is the stochastic load vector with noise intensity matrix $\boldsymbol{\Sigma}$. $\boldsymbol{\alpha}$ in Equation (9.1) represents the parameters corresponding to the nonlinear stiffness model. Note that $\mathbf{M}_0$, $\mathbf{C}_0$, and $\mathbf{K}_0$ are the nominal parameters and represents the pristine system.

9.1.2 THE DIGITAL TWIN

The DT for the N-DOF nonlinear system discussed above can be expressed as:

$$\mathbf{M}(t_s)\frac{\partial^2 \boldsymbol{X}(t,t_s)}{\partial t^2} + \mathbf{C}(t_s)\frac{\partial \boldsymbol{X}(t,t_s)}{\partial t} + \mathbf{K}(t_s)\boldsymbol{X}(t,t_s) + \boldsymbol{G}\left((t,t_s),\boldsymbol{\alpha}\right)$$
$$= \boldsymbol{F}(t,t_s) + \boldsymbol{\Sigma}\dot{\boldsymbol{W}} \tag{9.2}$$

DOI: 10.1201/9781003268048-9

where t represents the system's time and t_s is the service time (operational time-scale). The response vector X is a function of both the time-scales and therefore, partial derivatives have been used in Equation (9.2). Equation (9.2) is considered to be the DT for the nominal system in Section 9.1.1. Equation (9.2) has two time-scales, t and t_s. For all practical purposes the service time-scale t_s is much slower as compared to the time-scale of the system dynamics.

9.1.3 PROBLEM STATEMENT

Although a physics-based DT for MDOF nonlinear system is defined in Equation (9.2), to use it in practice, one needs to estimate the system parameters $\mathbf{M}(t_s)$, $\mathbf{C}(t_s)$ and $\mathbf{K}(t_s)$. For estimating these parameters, the connectivity between the physical twin and the DT is crucial. Recent developments in IoT provides several new technologies that ensure the connectivity between the two twins. To be specific, the two-way connectivity between the DT and its counterpart is created by using sensors and actuators. Given the huge difference in the two times-scales in Equation (9.2), it is reasonable to assume that the temporal variation in $\mathbf{M}(t_s)$, $\mathbf{C}(t_s)$ and $\mathbf{K}(t_s)$ are so slow that the dynamics is practically decoupled from these parametric variations. The sensor collects data intermittently at discrete time instants t_s. At each time point t_s, time history measurements of acceleration response in $t_s \pm \Delta t$ is available. For this study, it is assumed that there is no practical variation in the mass matrix and hence, $\mathbf{M}(t_s) = \mathbf{M}_0$. Variation in damping matrix is also not addressed. With this setup, the objective is to develop a DT for nonlinear MDOF system. It is envisioned that the DT will track the variation in the system parameters, $\mathbf{K}(t_s)$ at current time t and is also able to predict future degradation/variation in system parameters. Last but not the least, a DT should be continuously updated as and when it receives data.

9.2 BAYESIAN FILTERING ALGORITHM

A crucial component in development of DT is estimating $\mathbf{K}(t_s)$ given the observations until time t_s. This is a classical parameter estimation problem and this chapter proposes the use of Bayesian filter to accomplish the goal. However, one must realize that development of DT and parameter estimation are not same; instead, parameter estimation is only a subset of the overall DT.

Bayesian filters use Bayesian inference to develop a framework which can then be applied for state-parameter estimation. Bayesian inference differs from conventional frequentist approach of statistical inference because it takes probability of an event as the uncertainty of the event in a single trial, as opposed to the proportion of the event in a probability space. For filtering equations, let the unknown vector be given as $\boldsymbol{Y}_{0:T} = \{\boldsymbol{Y}_0, \boldsymbol{Y}_1, \ldots, \boldsymbol{Y}_T\}$ which is observed through a set of noisy measurements $\boldsymbol{Z}_{1:T} = \{\boldsymbol{Z}_1, \boldsymbol{Z}_2, \ldots, \boldsymbol{Z}_T\}$.

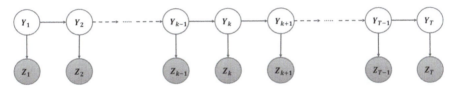

Figure 9.1 Probabilistic graphical model for state space model. We have considered first order Markovian assumption for the hidden variable $\boldsymbol{Y}$; this ensures that $\boldsymbol{Y}_t$ is only dependent on $\boldsymbol{Y}_{t-1}$.

Using Bayes's rule,

$$p(\boldsymbol{Y}_{0:T}|\boldsymbol{Z}_{1:T}) = \frac{p(\boldsymbol{Z}_{1:T}|\boldsymbol{Y}_{0:T})p(\boldsymbol{Y}_{0:T})}{p(\boldsymbol{Z}_{1:T})}. \tag{9.3}$$

This full posterior formulation, although accurate, is computationally onerous and is often intractable. The computational complexity is ameliorated by using the first-order Markovian assumption. First-order Markov model assumes (i) the state of system at time step k (i.e., $\boldsymbol{Y}_k$) given the state at time step $k-1$ (i.e., $\boldsymbol{Y}_{k-1}$) is independent of anything that has happened before time step $k-1$ and (ii) The measurement at time step k (i.e., $\boldsymbol{Z}_k$) given the state at time step k (i.e., $\boldsymbol{Y}_k$) is independent of any measurement or state histories. Mathematically, this is represented as:

$$p(\boldsymbol{Y}_k|\boldsymbol{Y}_{1:k-1}, \boldsymbol{Z}_{1:k-1}) = p(\boldsymbol{Y}_k|\boldsymbol{Y}_{k-1}) \tag{9.4}$$

and

$$p(\boldsymbol{Z}_k|\boldsymbol{Y}_{1:k}, \boldsymbol{Z}_{1:k-1}) = p(\boldsymbol{Z}_k|\boldsymbol{Y}_k) \tag{9.5}$$

A probabilistic graphical model representing the first-order Markov assumption is shown in Figure 9.1. In literature, this is also known as the state-space model (if the state is continuous) or the hidden Markov model (if the state is discrete). Using assumptions of Markovian model, the recursive Bayesian filter can be set up, and Kalman Filter arises [170, 205], which is a special case of recursive Bayesian filter used for linear models. Extended Kalman Filter [170] and Unscented Kalman Filter (UKF) [170, 189] are improvements over the Kalman filter, which are used for nonlinear models. In this chapter, UKF is used as the Bayesian filtering algorithm of choice. UKF is computationally expensive as compared to the EKF algorithm; however, the performance of UKF for systems having higher order of nonlinearity is superior [189].

9.2.1 UNSCENTED KALMAN FILTER

UKF uses concepts of unscented transform to analyze nonlinear models and approximates the mean and co-variance of the targeted distribution instead of approximating the nonlinear function. To that end, weighted sigma points

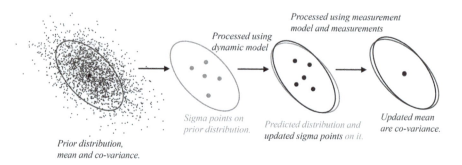

Figure 9.2 Schematic representation of functionality of sigma points within the UKF framework.

are used. The idea here is to consider some points on the source Gaussian distribution which are then mapped onto the target Gaussian distribution after passing through nonlinear function. These points are referred to as the sigma points and are considered to be representative of the transformed Gaussian distribution. Considering L to be the length of the state vector, $2L + 1$ sigma points are selected as [189],

$$\mathcal{Y}^{(0)} = \boldsymbol{\mu}$$

$$\mathcal{Y}^{(i)} = \boldsymbol{\mu} + \sqrt{(L + \lambda)} \left[\sqrt{\boldsymbol{\Sigma}} \right], \qquad i = 1, \ldots, L \qquad (9.6)$$

$$\mathcal{Y}^{(i)} = \boldsymbol{\mu} - \sqrt{(L + \lambda)} \left[\sqrt{\boldsymbol{\Sigma}} \right] \qquad i = L + 1, \ldots, 2L$$

where, $\mathcal{Y}$ are the required sigma points, $\boldsymbol{\mu}$ and $\boldsymbol{\Sigma}$ are, respectively the mean vector and co-variance matrix. λ and L in Equation (9.6) represent the scaling parameter and length of state vector, respectively. Details on how to compute λ are provided while discussing the UKF algorithm. Once the mean m_k and covariance p_k, are computed using the UKF, the filtering distribution is approximated as:

$$p(y_k|z_{1:k}) \simeq N(y_k|m_k, p_k) \qquad (9.7)$$

where m_k and p_k are the mean and co-variance computed by the algorithm discussed next. A schematic representation explaining how sigma points are used within the UKF algorithm is shown in Figure 9.2

9.2.1.1 Algorithm

Step 1 Weights calculation for sigma points

Select UKF parameters : $\alpha_f = 0.001$, $\beta = 2$, $\kappa = 0$

$$W_m^{(i=0)} = \frac{\lambda}{L + \lambda}$$

$$W_c^{(i=0)} = \frac{\lambda}{L + \lambda} + (1 - \alpha_f^2 + \beta)$$

$$W_m^{(i)} = \frac{1}{2(L + \lambda)}, \qquad\qquad i = 1, ..., 2L$$

$$W_c^{(i)} = W_m^{(i)}, \qquad\qquad i = 1, ..., 2L$$

(9.8)

where, L is the length of state vector and scaling parameter $\lambda = \alpha_f^2(L + \kappa) - L$.

Step 2 : For k = 0

Initialize mean and co-variance i.e., $m_k = m_0$, $p_k = p_0$.

Step 3 : For k = 1,2,.....,t_n

Step 3.1 : Prediction

Getting Sigma points $\mathcal{Y}^{(i)}, i = 0, ..., 2L$

$$\mathcal{Y}_{k-1}^{(0)} = m_{k-1}$$

$$\mathcal{Y}_{k-1}^{(i)} = m_{k-1} + \sqrt{L + \lambda} \left[\sqrt{P_{k-1}} \right]$$

$$\mathcal{Y}_{k-1}^{(i+L)} = m_{k-1} - \sqrt{L + \lambda} \left[\sqrt{P_{k-1}} \right], \qquad i = 1, ..., L.$$

(9.9)

Propagate sigma points through the dynamic model

$$\mathcal{Y}_k^{(i)} = f(\mathcal{Y}_{k-1}^{(i)}), \qquad i = 0, ..., 2L.$$

(9.10)

The predicted mean m_k^- and co-variance P_k^- is then given by

$$m_k^- = \sum_{i=0}^{2L} W_m^{(i)} \mathcal{Y}_k^{(i)},$$

$$P_k^- = \sum_{i=0}^{2L} W_c^{(i)} (\mathcal{Y}_k^{(i)} - m_k^-)(\mathcal{Y}_k^{(i)} - m_k^-)^T + Q_{k-1}.$$

(9.11)

Step 3.2 : Update

Getting Sigma points

$$\mathcal{Y}_k^{-(0)} = m_k^-$$

$$\mathcal{Y}_k^{-(i)} = m_k^- + \sqrt{L + \lambda} \left[\sqrt{P_k^-} \right]$$

$$\mathcal{Y}_k^{-(i+L)} = m_k^- - \sqrt{L + \lambda} \left[\sqrt{P_k^-} \right], \qquad i = 1, ..., L.$$

(9.12)

Propagating sigma points through the measurement model

$$\mathcal{Z}_k^{(i)} = h(\mathcal{Y}_k^{-(i)}), \qquad i = 0, \ldots, 2L. \tag{9.13}$$

Getting mean μ_k, predicted co-variance S_k and cross co-variance C_k

$$\mu_k^- = \sum_{i=0}^{2L} W_m^{(i)} \mathcal{Z}_k^{(i)},$$

$$S_k^- = \sum_{i=0}^{2L} W_c^{(i)} (\mathcal{Z}_k^{(i)} - \mu_k)(\mathcal{Z}_k^{(i)} - \mu_k)^T + R_k, \tag{9.14}$$

$$C_k^- = \sum_{i=0}^{2L} W_c^{(i)} (\mathcal{Y}_k^{-(i)} - m_k^-)(\mathcal{Z}_k^{(i)} - \mu_k)^T.$$

Step 3.3 : Getting filter gain K_k, filtered state mean m_k and co-variance P_k

conditional on measurement y_k.

$$K_k = C_k S_k^{-1},$$
$$m_k = m_k^- + K_k[y_k - \mu_k], \tag{9.15}$$
$$P_k = P_k^- - K_k S_k K_k^T.$$

Within the DT framework, the UKF algorithm is used for parameter estimation at a given time-step, t_k.

9.3 SUPERVISED MACHINE LEARNING ALGORITHM

In this section, we highlight the other component of the proposed DT framework, namely Gaussian process regression (GPR). GPR [18, 135], along with neural network [27, 108] are perhaps the most popular machine learning techniques in today's time. Unlike conventional frequentist machine learning techniques, GPR doesn't assume a functional form to represent input-output mapping; instead, a distribution over a function is assumed in GPR. Consequently, GPR has the inherent capability of capturing the epistemic uncertainty [96] arising due to limited data. This feature of GPR is particularly useful when it comes to decision making. Within the proposed DT framework, we use GPR to track the temporal evolution of the system parameters.

We consider $\boldsymbol{v}_k$ to be the systems parameters and time τ_k. In GPR, we represent $\boldsymbol{v}_k$ as

$$\boldsymbol{v}_k \sim \mathcal{GP}\left(\boldsymbol{\mu}(\tau_k; \boldsymbol{\beta}), \boldsymbol{\kappa}(\tau_k, \tau_k'; \sigma^2, \boldsymbol{l})\right) \tag{9.16}$$

where $\boldsymbol{\mu}(\cdot; \boldsymbol{\beta})$ and $\boldsymbol{\kappa}(\cdot, \cdot; \sigma^2, \boldsymbol{l})$, respectively represent the mean function and the covariance function of the GPR. The mean function is parameterized by the unknown coefficient vector $\boldsymbol{\beta}$ and the covariance function is parameterized by the process variance σ^2 and the length-scale parameters $\boldsymbol{l}$. All the parameters combined, $\boldsymbol{\theta} = [\boldsymbol{\beta}, \boldsymbol{l}, \sigma^2]$ are known as hyperparamters of GPR. It is worthwhile to note that choice of $\boldsymbol{\mu}(\cdot; \boldsymbol{\beta})$ and $\boldsymbol{\kappa}(\cdot, \cdot; \sigma^2, \boldsymbol{l})$ has significant influence on the performance of GP; this naturally allows an user to encode prior knowledge into the GPR model and model complex functions [135]. In case there is no prior knowledge about the mean function, it is a common practice to use zero mean Gaussian process,

$$\boldsymbol{v}_k \sim \mathcal{GP}\left(\mathbf{0}, \boldsymbol{\kappa}(\tau_k, \tau_k'; \sigma^2, \boldsymbol{l})\right) \tag{9.17}$$

The covariance function $\kappa(\cdot, \cdot; \sigma^2, \mathbf{1})$, on the other hand, should result in a positive, semi-definite matrix. For using the GPR in practice, one needs to compute the hyperparameters $\boldsymbol{\theta}$ based on training samples $\mathcal{D} = [\tau_k, \boldsymbol{v}_k]_{k=1}^{N_s}$ where N_s is the number of training samples. The most widely used method in this regard is based on the maximum likelihood estimation where the negative log-likelihood of GPR is minimized. For details on MLE for GPR, interested readers may refer [155]. The other alternative is to compute the posterior distribution of hyperparameter vector $\boldsymbol{\theta}$ [18, 19]. Though this is a superior alternative, it renders the process computationally expensive. In this chapter, we have used the MLE-based approach because of this simplicity. For the benefit of readers, the steps involved in training a GPR model are shown in Algorithm 6.

Once the hyper-parameters $\boldsymbol{\theta}$ are obtained, predictive mean and predictive variance corresponding to new input τ^* are computed as

$$\boldsymbol{\mu}^* = \boldsymbol{\Phi}\boldsymbol{\beta}^* + \boldsymbol{\kappa}^*(\tau^*; (\sigma^*)2, \boldsymbol{l}^*)\mathbf{K}^{-1}\left(\boldsymbol{v} - \boldsymbol{\Phi}\boldsymbol{\beta}^*\right), \tag{9.18}$$

$$s^2(\tau^*) = (\sigma^*)^2 \left\{ 1 - \boldsymbol{\kappa}^*(\tau^*; (\sigma^*)2, \boldsymbol{\theta}^*)\mathbf{K}^{-1}\boldsymbol{\kappa}^*(\tau^*; (\sigma^*)2, \boldsymbol{l}^*)^T \right.$$
$$\left. + \frac{\left[1 - \boldsymbol{\Phi}^T\mathbf{K}^{-1}\boldsymbol{\kappa}^*(\tau^*; (\sigma^*)2, \boldsymbol{l}^*)^T\right]}{\boldsymbol{\Phi}^T\mathbf{K}^{-1}\boldsymbol{\Phi}} \right\} \tag{9.19}$$

where $\boldsymbol{\beta}^*$, $\boldsymbol{l}^*$, and σ^* represents the optimized hyper-parameters. $\boldsymbol{\Phi}$ in Equations (9.18) and (9.19) represents the design matrix. $\boldsymbol{\kappa}^*(\tau^*; (\sigma^*)2, \boldsymbol{l}^*)^T$ in Equations (9.18) and (9.19) are the covariance vector between the input training samples and τ^* and computed as

$$\boldsymbol{\kappa}^*(\tau^*; (\sigma^*)2, \boldsymbol{l}^*)^T = [\kappa(\tau^*, \tau_1; ; (\sigma^*)2, \boldsymbol{\theta}^*), \dots, \kappa(\tau^*, \tau_{N_s}; ; (\sigma^*)2, \boldsymbol{\theta}^*)] \tag{9.20}$$

9.4 HIGH FIDELITY PREDICTIVE MODEL

We have discussed UKF and GP, the two ingredients of the proposed approach in this chapter. We proceed to discuss the proposed DT framework for nonlinear dynamical systems. A schematic representation of the proposed

---

Algorithm 6 Training GPR

Pre-requisite: Form of mean function $\boldsymbol{\mu}(\cdot;\boldsymbol{\beta})$ and covariance function, $\kappa(\cdot,\cdot;\sigma^2,\mathbf{l})$. Provide training data, $\mathcal{D} = [\tau_k, \boldsymbol{v}_k]_{k=1}^{N_s}$, initial values of the parameters, $\boldsymbol{\theta}_0$, maximum allowable iteration n_{max} and error threshold ϵ_t.
$\boldsymbol{\theta} \leftarrow \boldsymbol{\theta}_0$; $iter \leftarrow 0$; $\epsilon \leftarrow 10\epsilon_t$

repeat $iter \geq n_{max}$ and $\epsilon > \epsilon_t$
$iter \leftarrow iter + 1$
$\boldsymbol{\theta}_{iter-1} \leftarrow \boldsymbol{\theta}$.

Compute the negative log-likelihood by using the training data $\mathcal{D}$ and $\boldsymbol{\theta}$

$$f_{ML} \propto \frac{1}{N}\left|\mathbf{K}(\boldsymbol{\theta}) + \log\left(\boldsymbol{v}^T\mathbf{R}(\boldsymbol{\theta})^{-1}\boldsymbol{v}\right)\right|,$$

where $\mathbf{K}(\boldsymbol{\theta})$ is the covariance matrix computed by using the training data and covariance function $\kappa(\cdot,\cdot;\sigma^2,\mathbf{l})$. $\boldsymbol{v}$ represents the observation vector. Update hyperparameter $\boldsymbol{\theta}$ based on the gradient information.

$\boldsymbol{\theta}_{iter} \leftarrow \boldsymbol{\theta}$.

$\epsilon \leftarrow \|\boldsymbol{\theta}_{iter} - \boldsymbol{\theta}_{iter-1}\|_2^2$ **Output:** Optimal hyper-parameter, $\boldsymbol{\theta}^*$

DT is shown in Figure 9.3. It has four primary components, namely (a) selection of nominal model, (b) data collection, (c) parameter estimation at a given time-instant, and (d) estimation of the temporal variation in parameters. The selection of nominal model has already been detailed earlier and hence, here the discussion is limited to data collection, parameter estimation, and estimation of temporal variation of the parameters only.

A major concern in DT is its connectivity with the physical twin; in absence of which, a DT will be of no practical use. To ensure connectivity, sensors are placed on the physical system (physical twin) for data collection. The data is communicated to the DT by using cloud technology. With the substantial advancements in IoT, the access to different types of sensors is straightforward for collecting different types of data. In this chapter, we have considered that accelerometers are mounted on the physical system and the DT receives acceleration measurements. To be specific, one can consider that the acceleration time-history are available to the DT intermittently at discrete time-instant t_s. Note that the proposed approach is equally applicable (with trivial modifications) if instead of acceleration, displacement or velocity measurements are available. The framework can also be extended to function in tandem with vision based sensors. However, from a practical and economic

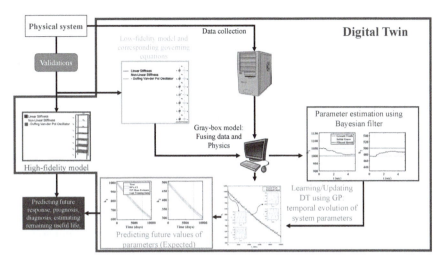

Figure 9.3 Schematic representation of the proposed digital twin framework. It comprises of low-fidelity model as nominal model, UKF for parameter estimation, GP for learning temporal evolution of parameters and predicting future values of system parameters, and high-fidelity model for estimating future responses.

point-of-view, it is easiest to collect acceleration measurements and hence, the same has been considered in this study.

Once the data is collected, the next objective is to estimate the system parameters (stiffness matrix to be specific), assuming that at time-instant t_s, acceleration measurements are avilable in $[t_s - \Delta t, t_s]$, where Δt is the time interval over which acceleration measurement is available at t_s. Here t_s is a time-step in the slow time-scale whereas Δt is time interval in the fast time-scale. With this setup, the parameter estimation objective is to estimate $\mathbf{K}(t_s)$. In this chapter, we estimate $\mathbf{K}(t_s)$ by using the UKF.

The last step within the proposed DT framework is to estimate the temporal evolution of the parameters. This is crucial as it enables the DT to predict future behavior of the physical system. In this chapter, we propose to use a combination of GPR and UKF for learning the temporal evolution of the system parameters. To be specific, consider $\mathbf{t} = [t_1, t_2, \ldots, t^N]$ to be time-instants in slow-scale. Furthermore, assume that using UKF, the estimated system parameters are available at different time-instants as $\mathbf{v} = [\mathbf{v}_1, \mathbf{v}_2, \ldots, \mathbf{v}_N]$, where $\mathbf{v}_i$ includes the elements of stiffness matrix. The proposed work trains a GPR model between $\mathbf{t}$ and $\mathbf{v}$,

$$\mathbf{v} \sim \mathcal{GP}(\boldsymbol{\mu}, \boldsymbol{\kappa}) \tag{9.21}$$

For brevity, the hyperparameters in Equation (9.21) are omitted. The GPR is trained by following the procedure discussed in Algorithm 6. Once trained, the GPR can predict the system parameters at future time-steps. Since GPR is a

Bayesian machine learning model it also provides the predictive uncertainty which can be used to judge the accuracy of the model. For the ease of readers, the overall DT framework proposed is shown in Algorithm 7.

Algorithm 7 Proposed DT

Select nominal model

▷ Section 9.1.

Use data (acceleration measurements) $\mathcal{D}_s$ collected at time t_s to compute the parameters $\mathbf{K}(t_s)$

▷ Section 9.2.

Train a GP using $\mathcal{D} = [t_n, \boldsymbol{v}_n]_{n=1}^{t_s}$ as training data, where $\boldsymbol{v}_n$ represents the system parameter

▷ Algorithm 6. Predict $\mathbf{K}(\tilde{t})$ at future time $\tilde{t}$

▷ Using trained mixture of Gaussian processes.

Substitute $\mathbf{K}(\tilde{t})$ into the governing equation (high-fidelity model) and solve it to obtain responses at time $\tilde{t}$.

Take decisions related to maintenance, remaining useful life and health of the system.

Repeat steps $2-6$ as more data become available

9.5 EXAMPLES

In this section, we present two examples to illustrate the performance of the proposed DT framework.

1. The first example is a 2-DOF system with duffing oscillator attached at the first floor.
2. The second example is a 7-DOF system. For this example, the nonlinearity in the system arises because of a duffing van der pol oscillator attached between the third and the fourth DOF.

As stated earlier, we have considered that acceleration measurements at different time-steps are available. The objective here is to use the proposed DT to compute the time-evolution of the system parameters. Once the time-evolution of the parameters are known, the proposed DT can be used for predicting the responses of the system at future time-steps (t_s) (see Algorithm 7 for details). In this section, we have illustrated how the proposed approach can be used for predicting the time-evolution of the system parameters in the past as well as in the future.

9.5.1 2-DOF SYSTEM WITH DUFFING OSCILLATOR

As the first example, we consider a 2-DOF system as shown in Figure 9.4. The nonlinear duffing oscillator is attached with the first degree of freedom. The coupled governing equations for this system are represented as

$$m_1 \ddot{x}_1 + c_1 \dot{x}_1 + k_1 x_1 + \alpha_{DO} x_1^3 + c_2(\dot{x}_1 - \dot{x}_2) + k_2(x_1 - x_2) = \sigma_1 \dot{W}_1 + f_1$$

$$m_2 \ddot{x}_2 + c_2(\dot{x}_2 - \dot{x}_1) + k_2(x_2 - x_1) = \sigma_2 \dot{W}_2 + f_2$$

$$(9.22)$$

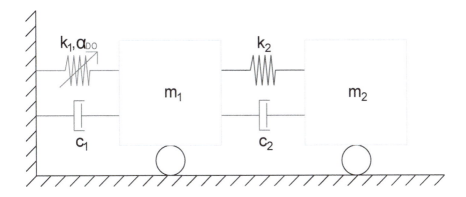

Figure 9.4 Schematic representation of the 2-DOF System with duffing oscillator considered in example 1. The nonlinear duffing oscillator is attached with the first degree of freedom (shown in magenta).

where m_i, c_i, and k_i, respectively, represent the mass, damping, and stiffness of the ith degree of freedom. Although not explicitly shown, it is to be noted that k_i changes with the slow time-scale t_s. F_i and $\sigma_i \dot{W}_i$, respectively, represents the deterministic and the stochastic force acting on the ith floor. α_{DO} controls the nonlinearity in the system. The parametric values considered for this example are shown in Table 9.1.

The system states are defined as:

$$x_1 = y_1, \quad x_2 = y_2,$$
$$\dot{x}_1 = y_3, \quad \dot{x}_2 = y_4 \tag{9.23}$$

and the governing equation in Equation (9.22) is represented in the form of Ito-diffusion equations to obtain the drift and dispersion coefficients:

$$dy = a\,dt + b\,dW \tag{9.24}$$

Table 9.1

System Parameters for 2-DOF System

Mass (Kg)	Stiffness Constant (N/m)	Damping Constant (Ns/m)	Force(N) $F_i = \lambda_i sin(\omega_i t)$	Stochastic Noise Parameters
$m_1 = 20$	$k_1 = 1000$	$c_1 = 10$	$\lambda_1 = 10, \ \omega_1 = 10$	$s_1 = 0.1$
$m_2 = 10$	$k_2 = 500$	$c_2 = 5$	$\lambda_2 = 10, \ \omega_2 = 10$	$s_2 = 0.1$
DO Oscillator Constant, $\alpha_{DO} = 100$				

where

$$
\boldsymbol{a} = \begin{bmatrix} y_3 \\ y_4 \\ \frac{f_1}{m_1} - \frac{1}{m_1}\left(c_1\,y_3 + c_2\,y_3 - c_2\,y_4 + k_1\,y_1 + k_2\,y_1 - k_2\,y_2 + \alpha_{do}\,y_1{}^3\right) \\ \frac{1}{m_2}\left(c_2\,y_3 - c_2\,y_4 + k_2\,y_1 - k_2\,y_2 m_2\right) + \frac{f_2}{m_2} \end{bmatrix}
$$

(9.25a)

$$
\mathbf{b} = \begin{bmatrix} 0 & 0 \\ 0 & 0 \\ \frac{\sigma_1}{m_1} & 0 \\ 0 & \frac{\sigma_2}{m_2} \end{bmatrix}
$$

(9.25b)

To illustrate the performance of the proposed digital twin, we generate synthetic data by simulating Equation (9.24). The data simulation is carried out using Taylor 1.5 strong scheme [161, 185]

$$
\begin{aligned}
\boldsymbol{y}_{k+1} = \big(\boldsymbol{y} + \boldsymbol{a}\Delta t + \mathbf{b}\Delta w &+ 0.5 L^j(\mathbf{b})\left(\Delta w^2 - \Delta t\right) + L^j(\boldsymbol{a})\Delta z \\
&+ L^0(\mathbf{b})\left(\Delta w \Delta t - \Delta z\right) + 0.5 L^0(\boldsymbol{a})\Delta t^2\big)_k
\end{aligned}
$$

(9.26)

where, L^0 and L^j are Kolmogorov operators [161] evaluated on drift and diffusion coefficients i.e., on elements of $\boldsymbol{a}$ and $\mathbf{b}$. Δw and Δz are the Brownian increments [161] evaluated at each time step Δt. Before proceeding with the performance of the proposed DT, we investigate the performance of UKF in joint parameter state estimation. To avoid the so called inverse crime [209], Euler Maruyama (EM) integration scheme is used during filtering.

$$
\boldsymbol{y}_{k+1} = \left(\boldsymbol{y} + \boldsymbol{a}\Delta t + \mathbf{b}\Delta w\right)_k
$$

(9.27)

EM integration scheme provides a lower-order approximation as compared to Taylor's 1.5 strong integration scheme. In other words, the data is generated using a more accurate scheme as compared to the filtering. This helps in emulating a realistic scenario. For combined state-parameter estimation, the state space vector is modified as $\boldsymbol{y} = [y_1\ y_2\ y_3\ y_4\ k_1\ k_2]^T$. Consequently, $\boldsymbol{a}$ and $\mathbf{b}$ are also modified as:

$$
\boldsymbol{a} = \begin{bmatrix} y_3 \\ y_4 \\ \frac{f_1}{m_1} - \frac{1}{m_1}\left(c_1\,y_3 + c_2\,y_3 - c_2\,y_4 + k_1\,y_1 + k_2\,y_1 - k_2\,y_2 + \alpha_{do}\,y_1{}^3\right) \\ \frac{1}{m_2}\left(c_2\,y_3 - c_2\,y_4 + k_2\,y_1 - k_2\,y_2\right) + \frac{f_2}{m_2} \\ 0 \\ 0 \end{bmatrix}
$$

(9.28a)

$$\mathbf{b} = \begin{bmatrix} 0 & 0 \\ 0 & 0 \\ \frac{\sigma_1}{m_1} & 0 \\ 0 & \frac{\sigma_2}{m_2} \\ 0 & 0 \\ 0 & 0 \end{bmatrix} \tag{9.28b}$$

For obtaining the dynamic model function for UKF model, first two terms of EM algorithm are used.

$$\boldsymbol{f}(\boldsymbol{y}) = \boldsymbol{y} + \boldsymbol{a}\Delta t \tag{9.29}$$

For estimating the noise covariance $\mathbf{Q}$, q is expressed as:

$$\boldsymbol{q} = \mathbf{q_c}\boldsymbol{RV} \tag{9.30}$$

where $\mathbf{q}_c$ is a constant diagonal matrix which is multiplied by vector of random variables $\boldsymbol{RV}$ to compute $\boldsymbol{q}$. The basic form for $\mathbf{q}_c$ is extracted from the remaining terms of EM algorithm i.e., $\mathbf{b}\Delta\boldsymbol{w}$.

$$\mathbf{q_c} = diag \begin{bmatrix} 0 & 0 & \frac{\sigma_1\sqrt{dt}}{m_1} & \frac{\sigma_2\sqrt{dt}}{m_2} & 0 & 0 \end{bmatrix} \tag{9.31}$$

$$\mathbf{Q} = \mathbf{q_c}\mathbf{q}_c^T$$

The individual terms of $\mathbf{Q}$ can then be modified by any suitable factor to improve the accuracy of the filter. Since the acceleration measurements are available to the DT, the simulated acceleration measurements are obtained as follows:

$$\boldsymbol{A} = -\mathbf{M}^{-1}(\boldsymbol{G} + \mathbf{K}X + \mathbf{C}\dot{X}) \tag{9.32}$$

where $\mathbf{M}$, $\mathbf{C}$, and $\mathbf{K}$ are the mass, damping and stiffness matrices. $\boldsymbol{G}$, as already discussed in Equation (9.1) is the contribution due to the nonlinearity in the system. Equation (9.32) can be written in the state-space form as

$$\boldsymbol{A} = \begin{bmatrix} -\frac{1}{m_1}\left(c_1\,y_3 + c_2\,y_3 - c_2\,y_4 + k_1\,y_1 + k_2\,y_1 - k_2\,y_2 + \alpha_{do}\,y_1{}^3\right) \\ \frac{1}{m_2}\left(c_2\,y_3 - c_2\,y_4 + k_2\,y_1 - k_2\,y_2\right) \end{bmatrix} \tag{9.33}$$

Using Equation 9.33, the observation/measurement model for the UKF can be written as

$$\boldsymbol{h}(\boldsymbol{y}) = \begin{bmatrix} -\frac{1}{m_1}\left(c_1\,y_3 + c_2\,y_3 - c_2\,y_4 + k_1\,y_1 + k_2\,y_1 - k_2\,y_2 + \alpha_{do}\,y_1{}^3\right) \\ \frac{1}{m_2}\left(c_2\,y_3 - c_2\,y_4 + k_2\,y_1 - k_2\,y_2\right) \end{bmatrix} \tag{9.34}$$

The simulated acceleration measurements are contaminated by white Gaussian noise having a signal-to-noise ratio (SNR) of 50, where SNR is defined as: $\text{SNR} = \sigma_{\text{signal}}^2/\sigma_{\text{noise}}^2$ and σ is the standard deviation. The deterministic force vector is also corrupted by white Gaussian noise having SNR of 20. Representative examples of acceleration and deterministic force for this problem are shown in Figure 9.5. We use UKF along with the acceleration

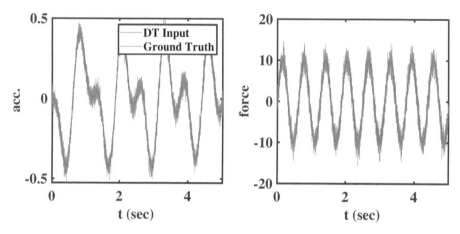

Figure 9.5 Sample Acceleration and deterministic component of the force for the 2-DOF problem. The stochasticity observed for the force is due to the presence of noise. Note that there is an additional stochastic component of force as shown in Equation (9.27).

and deterministic force measurements, $f(y)$ and $h(y)$ for combined parameter state estimation. Figure 9.7 shows the combined state parameter estimation results for first data point i.e., $t_{s(i)} = t_{s(1)}$ of 2-DOF system. This is a relatively straightforward case where measurements at both degrees of freedom are available (see Figure 9.6). It can be observed that UKF provides highly accurate estimates of the state vectors. As for parameter estimation (see Figure 9.7(b)), we observe that UKF provides highly accurate estimate for k_1. As for k_2, compared to the ground truth ($k_2 = 500\,\mathrm{N/m}$), the proposed approach ($k_2 = 487.5\,\mathrm{N/m}$) provides an accuracy of around 98%.

Figure 9.9 shows the result for an intermediate data point i.e., at time $t_{s(i)} = t_{s(91)}$. The initial values of parameter while filtering are taken as the final values of parameter obtained from previous data point. Similar to that observed for the initial data point, Figure 9.9 shows that the filter manages to estimate the states accurately and also improves upon the parameter estimation. Next, we consider a more complicated scenario where data at only one DOF is available. To be specific, acceleration measurements at DOF-1 is considered to be available (see Figure 9.10). This changes the measurement model $h(.)$ while filtering and reduces it to,

$$h(y) = -\frac{1}{m_1}\left(c_1\,y_3 + c_2\,y_3 - c_2\,y_4 + k_1\,y_1 + k_2\,y_1 - k_2\,y_2 + \alpha_{do}\,y_1{}^3\right) \quad (9.35)$$

Figure 9.11 shows the state and parameter estimation results for this case. Similar to previous case, it can be observed that the state estimate and the estimate for k_1 are obtained with high degree of accuracy (see Figure 9.11).

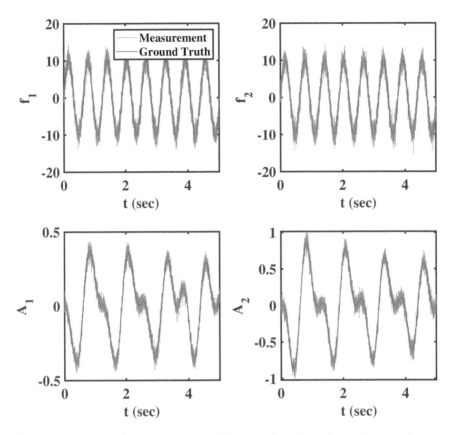

Figure 9.6 Deterministic component of force and acceleration vectors at the two DOFs. The noisy acceleration vectors are provided as measurement to the UKF model.

Estimate for k_2 also approaches the ground truth ($k_2 = 500\,\mathrm{N/m}$) giving an accuracy of approximately 98%. Finally, consider another objective of the DT, which is to compute the time-evolution of the parameters, considering the stiffness to vary with slow time-scale t_s as follows.

$$k(t_s) = k_0\delta \tag{9.36}$$

where

$$\delta = e^{-0.5\times10^{-4}\times t_s} \tag{9.37}$$

We consider that acceleration measurements are available for 5 seconds every 50 days. UKF is applied as discussed before for computing the stiffness at each time-steps. The resulting data obtained is shown in Figure 9.12. Once the data points are obtained, GP is employed to evaluate the temporal evolution of the parameters. Figure 9.13 shows results obtained using GP. The

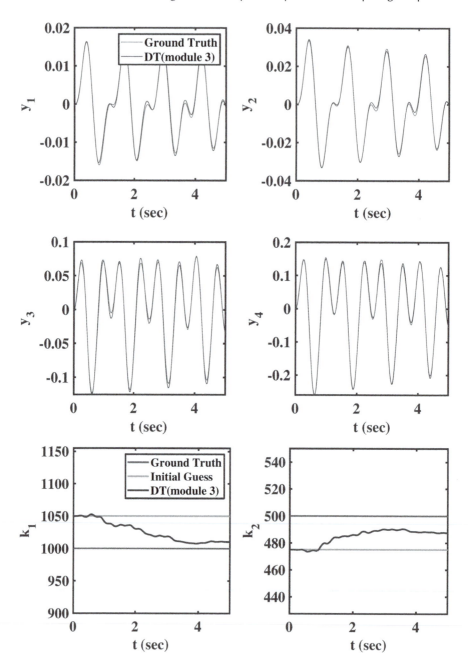

Figure 9.7 (a) State (Displacement and Velocity) Estimation and (b) Parameter (Stiffness) Estimation. Combined state and parameter estimation results for the 2-DOF system. Noisy measurements of acceleration at both the DOFs are provided as input to the UKF algorithm. The results corresponds to the initial measurement data.

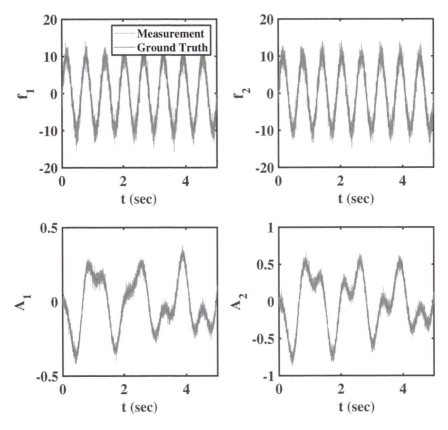

Figure 9.8 Force (deterministic part) and acceleration vectors at an intermediate time-step. The noisy accelerations at the 2-DOFs are provided as measurements to the UKF algorithm.

vertical lines in Figure 9.13 indicate the time until which data is provided to the GP. It is observed that GP yields a highly accurate estimate of the two stiffnesses. Interestingly, results obtained using GP are not only accurate in the time-window (indicated by the vertical line) but also outside. Therefore, the proposed DT can be used for predicting the system parameters at future time-step which in-turn can be used for predicting the future responses and solving remaining useful life and predictive maintenance optimization problems. Moreover, GP being a Bayesian machine learning algorithm provides an estimate of the confidence interval. This can be used for collecting more data and in decision making.

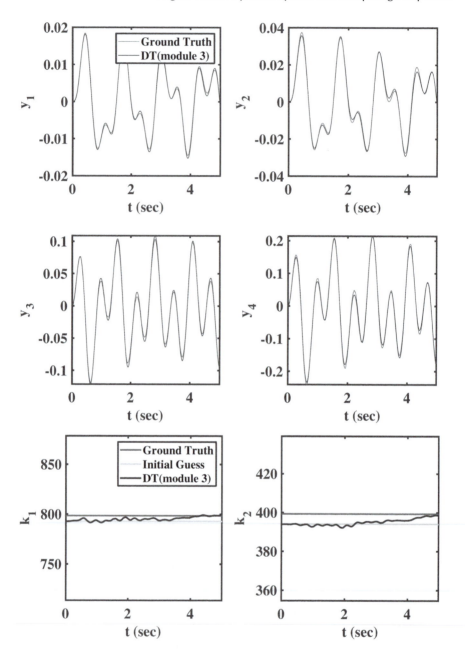

Figure 9.9 Combined state and parameter estimation results for the 2-DOF system. Noisy measurements of acceleration at both the DOFs are provided as input to the UKF algorithm. The results corresponds to an intermediate measurement data. The first two row represents state estimation while the third row represents parameter estimation.

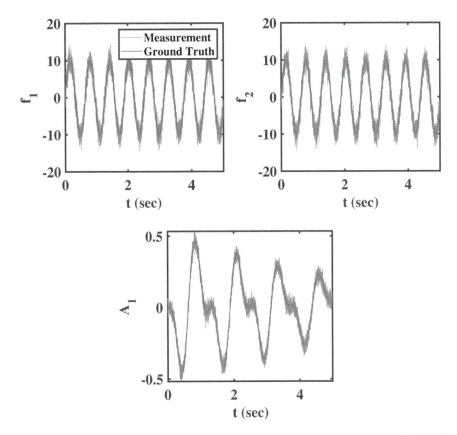

Figure 9.10 Deterministic component of force and acceleration vector used in UKF. For this case, only acceleration measurements at first degree of freedom is available.

9.5.2 7-DOF SYSTEM WITH DUFFING VAN DER POL OSCILLATOR

As our second example, we consider a 7-DOF system as shown in Figure 9.14. The 7-DOF system is modeled with a DVP oscillator at fourth DOF. The governing equations of motion for the 7-DOF system are given as,

$$\mathbf{M}\ddot{\mathbf{X}} + \mathbf{C}\dot{\mathbf{X}} + \mathbf{K}\mathbf{X} + \mathbf{G}(\mathbf{X}, \boldsymbol{\alpha}) = \mathbf{F} + \boldsymbol{\Sigma}\dot{\mathbf{W}} \tag{9.38}$$

where $\mathbf{M} = diag\,[m_1, \ldots, m_7] \in \mathbb{R}^{7\times 7}$, $\mathbf{X} = [x_1, \ldots, x_7]^T \in \mathbb{R}^7$, $\boldsymbol{\Sigma} = diag\,[\sigma_1, \ldots, \sigma_7] \in \mathbb{R}^{7\times 7}$, $\dot{\mathbf{W}} = \left[\dot{W}_1, \ldots, \dot{W}_7\right]^T \in \mathbb{R}^7$, $\mathbf{F} = [f_1, \ldots, f_7]^T \in \mathbb{R}^7$ and

$$\mathbf{G} = \alpha_{DVP} \begin{bmatrix} \mathbf{0}_{1\times 3} & (x_3 - x_4)^3 & (x_4 - x_3)^3 & \mathbf{0}_{1\times 2} \end{bmatrix}^T \in \mathbb{R}^7.$$

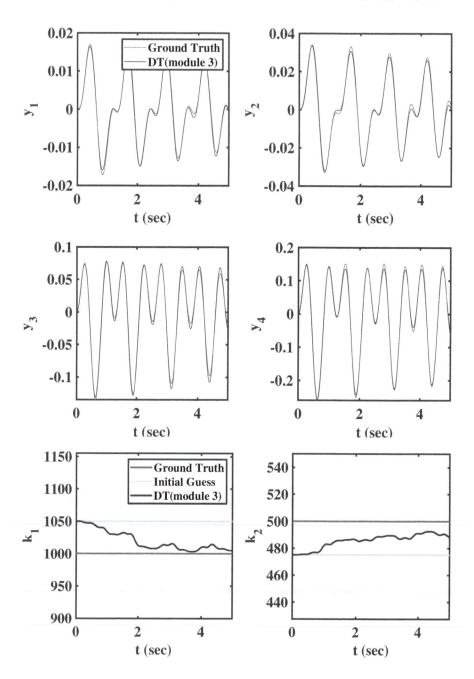

Figure 9.11 Combined state and parameter estimation results for the 2DOF system estimated from only one acceleration measurement. The noisy acceleration measurement at DOF 1 is provided to the UKF as measurement. The first two row represents state estimation while the third row represents parameter estimation.

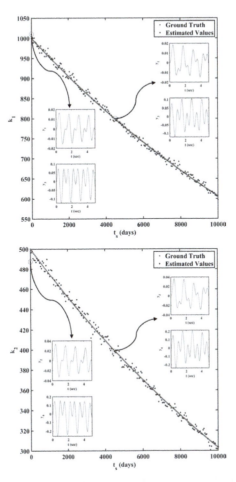

Figure 9.12 Estimated stiffness (k_1 and k_2) in slow-time-scale using the UKF algorithm for the 2-DOF example. State estimations at selected time-steps are also shown. Good match between the ground truth and the filtered result is obtained. These data act as input to the Gaussian process (GP).

$\mathbf{C}$ and $\mathbf{K}$ in Equation (9.38) are tri-diagonal matrices representing damping and stiffness (linear component),

$$\mathbf{C} = \begin{bmatrix} c_1 + c_2 & -c_2 \\ -c_2 & c_2 + c_3 & -c_3 \\ & -c_3 & c_3 + c_4 & -c_4 \\ & & -c_4 & c_4 + c_5 & -c_5 \\ & & & -c_5 & c_5 + c_6 & -c_6 \\ & & & & -c_6 & c_6 + c_7 & -c_7 \\ & & & & & -c_6 & c_7 \end{bmatrix} \quad (9.39)$$

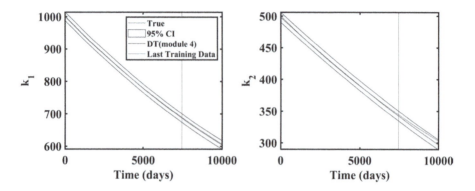

Figure 9.13 Results representing the performance of the proposed digital twin for the 2-DOF system. The GP is trained using the data generated using UKF. Data upto the horizontal line is available to the GP. The digital twin performs well even when predicting system parameters at future time-steps.

$$\mathbf{K} = \begin{bmatrix} k_1 + k_2 & -k_2 & & & & & \\ -k_2 & k_2 + k_3 & -k_3 & & & & \\ & -k_3 & k_3 - k_4 & k_4 & & & \\ & & k_4 & -k_4 + k_5 & -k_5 & & \\ & & & -k_5 & k_5 + k_6 & -k_6 & \\ & & & & -k_6 & k_6 + k_7 & -k_7 \\ & & & & & -k_6 & k_7 \end{bmatrix}$$

$$(9.40)$$

where m_i, c_i, and k_i, respectively, represent the mass, damping, and stiffness of the ith degree of freedom. We have considered the stiffness of all, but the fourth DOF to vary with the slow time-scale t_s. The rationale behind not varying the stiffness corresponding to the 4th DOF resides in the fact that nonlinear stiffness is generally used for vibration control [57] and energy harvesting [24] and hence, is kept constant. Parametric values for the 7-DOF system are shown in Table 9.2.

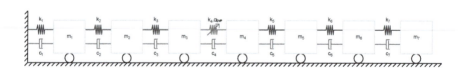

Figure 9.14 Schematic representation of the 7-DOF System with duffing Van-der Pol oscillator considered in example 2. The nonlinear DVP oscillator is attached with the fourth degree of freedom (shown in red).

Table 9.2
System Parameters − 7-DOF System − Data Simulation

Index	Mass	Stiffness	Damping	Force(N)	Stochastic Noise
i	(Kg)	Constant (N/m)	Constant (Ns/m)	$F_i = \lambda_i sin(\omega_i t)$	Parameters
$i = 1,2$	$m_i = 20$	$k_i = 2000$			
$i = 3,4,5,6$	$m_i = 10$	$k_i = 1000$	$c_i = 20$	$\lambda_i = 10,\ \omega_i = 10$	$s_i = 0.1$
$i = 7$	$m_i = 5$	$k_i = 500$			
		DVP Oscillator Constant, $\alpha_{DVP} = 100$			

To convert the governing equations for 7-DOF system to state space equations, the following transformations are considered:

$$x_1 = y_1, \quad \dot{x}_1 = y_2, \quad x_2 = y_3, \quad \dot{x}_2 = y_4, \quad x_3 = y_5, \quad \dot{x}_3 = y_6, \quad x_4 = y_7$$
$$\dot{x}_4 = y_8, \quad x_5 = y_9, \quad \dot{x}_5 = y_{10}, \quad x_6 = y_{11}, \quad \dot{x}_6 = y_{12}, \quad x_7 = y_{13}, \quad \dot{x}_7 = y_{14}$$
$$(9.41)$$

Using Equation (9.24), dispersion and drift matrices for 7-DOF system are identified as follows:

$$b_{ij} = \begin{cases} \dfrac{\sigma_i}{m_i}, & \text{for } i = 2j \text{ and } j = (1,2,3,5,6,7) \\ \dfrac{\sigma_i}{m_i} y_{2j-1}, & \text{for } i = 2j \text{ and } j = 4 \\ 0, & \text{elsewhere} \end{cases} \qquad (9.42)$$

$$a = \begin{bmatrix} y_2 \\ \frac{f_1}{m_1} - \frac{1}{m_1}(y_1(k_1+k_2) - c_2 y_4 - k_2 y_3 + y_2(c_1+c_2)) \\ y_4 \\ \frac{f_2}{m_2} + \frac{1}{m_2}(c_2 y_2 - y_3(k_2+k_3) + c_3 y_6 + k_2 y_1 + k_3 y_5 - y_4(c_2+c_3)) \\ y_6 \\ \frac{f_3}{m_3} - \frac{1}{m_3}(k_4 y_7 - c_4 y_8 - k_3 y_3 - c_3 y_4 + y_5(k_3-k_4) + \alpha_{DVP}(y_5-y_7)^3 + y_6(c_3+c_4)) \\ y_8 \\ \frac{f_4}{m_4} + \frac{1}{m_4}(c_4 y_6 + c_5 y_{10} - k_4 y_5 + k_5 y_9 + y_7(k_4-k_5) + \alpha_{DVP}\{y_5-y_7\}^3 - y_8(c_4+c_5)) \\ y_{10} \\ \frac{f_5}{m_5} + \frac{1}{m_5}(c_5 y_8 - y_9(k_5+k_6) + c_6 y_{12} + k_5 y_7 + k_6 y_{11} - y_{10}(c_5+c_6)) \\ y_{12} \\ \frac{f_6}{m_6} + \frac{1}{m_6}(c_6 y_{10} - y_{11}(k_6+k_7) + c_7 y_{14} + k_6 y_9 + k_7 y_{13} - y_{12}(c_6+c_7)) \\ y_{14} \\ \frac{f_7}{m_7} + \frac{1}{m_7}(c_7 y_{12} - c_7 y_{14} + k_7 y_{11} - k_7 y_{13}) \end{bmatrix}$$
$$(9.43)$$

Similar to previous example data simulation is carried out using Taylor-1.5-Strong algorithm shown in Equation (9.26) and filtering model is formed using EM equation shown in Equation (9.27). For performing combined state parameter estimation state vector is modified to:

$$y = [y_{1:14}, k_{1:7}]^T \qquad (9.44)$$

Consequently a and b matrices are changed as: $a = [a_{state}^T, \mathbf{0}_{1\times6}]^T$ and $b = [b_{state}^T, \mathbf{0}_{7\times7}]^T$ where a_{state} and b_{state} are equal to a and b from Equation (9.43) and Equation (9.42), respectively. Note that although the value k_4 is a-priori known, we have still considered it into the state vector. It was observed that such a setup helps in regularizing the UKF estimates. Dynamic

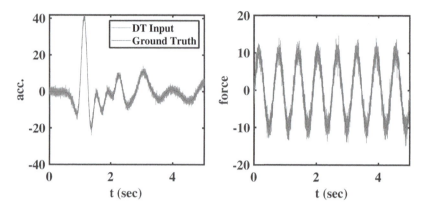

Figure 9.15 Sample Acceleration and deterministic component of the force for the 7-DOF problem. The stochasticity observed for the force is due to the noise present. Note that there is an additional stochastic component of force as shown in Equation (9.27).

model function, $\boldsymbol{f}(y)$ is obtained using Equation (9.29) and acceleration measurements are obtained using Equation (9.32). Since for measurement, accelerations of all DOF are considered, measurement model for the UKF remains same as acceleration model and can be written as,

$$
\boldsymbol{h}(\boldsymbol{y}) = \begin{bmatrix}
-\frac{1}{m_1}\left(y_1\ (k_1 + k_2) - c_2\ y_4 - k_2\ y_3 + y_2\ (c_1 + c_2)\right) \\
\frac{1}{m_2}\left(c_2\ y_2 - y_3\ (k_2 + k_3) + c_3\ y_6 + k_2\ y_1 + k_3\ y_5 - y_4\ (c_2 + c_3)\right) \\
-\frac{1}{m_3}\left(k_4\ y_7 - c_4\ y_8 - k_3\ y_3 - c_3\ y_4 + y_5\ (k_3 - k_4) + \alpha_{DVP}\ (y_5 - y_7)^3 + y_6\ (c_3 + c_4)\right) \\
\frac{1}{m_4}\left(c_4\ y_6 + c_5\ y_{10} - k_4\ y_5 + k_5\ y_9 + y_7\ (k_4 - k_5) + \alpha_{DVP}\ (y_5 - y_7)^3 - y_8\ (c_4 + c_5)\right) \\
\frac{1}{m_5}\left(c_5\ y_8 - y_9\ (k_5 + k_6) + c_6\ y_{12} + k_5\ y_7 + k_6\ y_{11} - y_{10}\ (c_5 + c_6)\right) \\
\frac{1}{m_6}\left(c_6\ y_{10} - y_{11}\ (k_6 + k_7) + c_7\ y_{14} + k_6\ y_9 + k_7\ y_{13} - y_{12}\ (c_6 + c_7)\right) \\
\frac{1}{m_7}\left(c_7\ y_{12} - c_7\ y_{14} + k_7\ y_{11} - k_7\ y_{13}\right)
\end{bmatrix} \quad (9.45)
$$

Process noise co-variance matrix $\mathbf{Q}$ is obtained using the same process as discussed for 2-DOF system (refer Equation [9.30]) and is written as,

$$
q_c = \sqrt{dt}\ diag\left[0\ \frac{\sigma_1}{m_1}\ 0\ \frac{\sigma_2}{m_2}\ 0\ \frac{\sigma_3}{m_3}\ 0\ \frac{m_k^-(7)\ \sigma_4}{m_4}\ 0\ \frac{\sigma_5}{m_5}\ 0\ \frac{\sigma_6}{m_6}\ 0\ \frac{\sigma_7}{m_7}\ 0\ 0\ 0\ 0\ 0\ 0\ 0\right]
$$

$$
\mathbf{Q} = q_c q_c^T
$$

$$(9.46)$$

Where, $m_k^-(7)$ is the seventh element of UKF's predicted mean calculated from Equation (9.11). The acceleration measurements and applied force are corrupted with a Gaussian noise having SNR values of 50 and 20 respectively. A comparison of the acceleration response obtained from data simulation and used in filtering is presented in Figure 9.15.

Similar to the previous example, we first investigate the performance of the UKF algorithm. To that end, the acceleration vectors (noisy) shown in Figure 9.16 is considered as the measurements. The state and parameter estimation results obtained using the UKF algorithm are shown in Figure 9.17. It can be observed that the proposed approach yields highly accurate estimate of

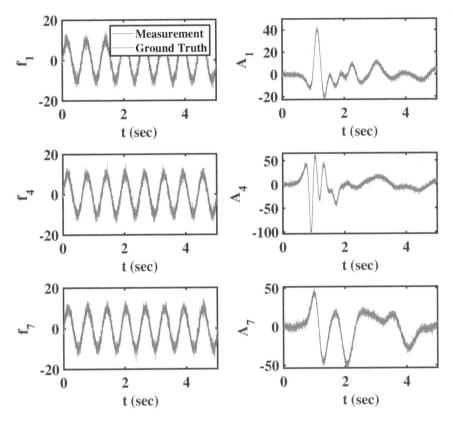

Figure 9.16 Deterministic component of force and acceleration vector corresponding to DOF 1, 4 and 7 used in UKF. The noisy acceleration vectors are provided as measurements to the UKF algorithm.

the state vectors. As for the parameter estimation, k_2, k_3, and k_5 converge exactly toward their respective true values. As for k_1, k_6, and k_7, UKF yields an accuracy of around 95%. A summary of the estimated parameters in the slow time-scales is shown in Figure 9.18 and Figure 9.19. We peruse that the estimates for new data points improve as our initial guess of system parameter improves (which for our case is the final parameters obtained from previous data points). Similar to Figure 9.17, we see that the estimates for stiffness k_2, k_3, and k_5 are more accurate than those obtained for k_1, k_6, and k_7. These data is used for training the GP model.

Figure 9.20 shows the results obtained using the GP. The vertical line in Figure 9.20 indicates the point until which data is available to the GP. For k_1, k_2, k_3, and k_5, the results obtained using GP matches exactly with the

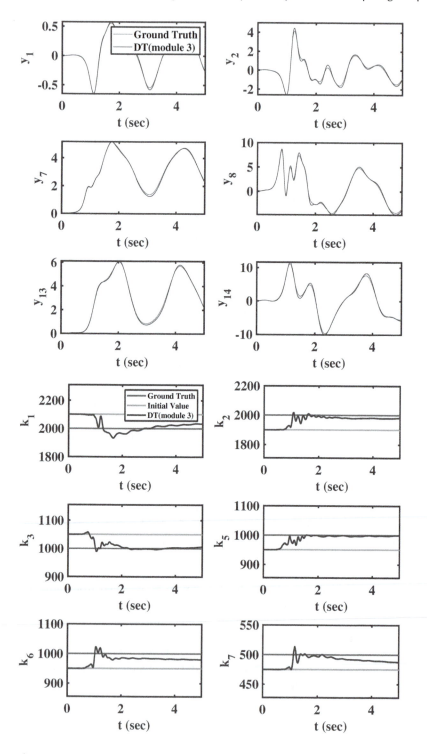

Figure 9.17 Combined state and parameter estimation results for the 7-DOF Van der Pol system.

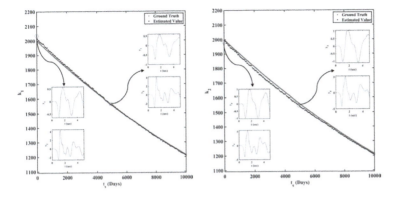

Figure 9.18 Estimated stiffness (k_1 and k_2) in slow-time-scale using the UKF algorithm for the 7-DOF example. State estimations at selected time-steps are also shown. Good match between the ground truth and the filtered result is obtained. These data act as input to the Gaussian process (GP).

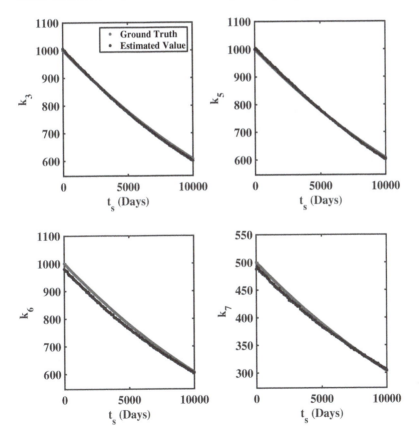

Figure 9.19 Estimated stiffness (k_3, k_5, k_6 and k_7) in slow-time-scale using the UKF algorithm for the 7-DOF example. Good match between the ground truth and the filtered result is obtained. These data act as input to the Gaussian process (GP).

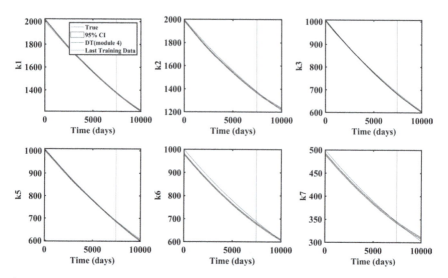

Figure 9.20 Results representing the performance of the proposed digital twin for the 7-DOF system. The GP is trained using the data generated using UKF. Data upto the horizontal line is available to the GP. The digital twin performs well even when predicting system parameters at future time-steps.

true solution. For k_7, the GP predicted results are found to diverge from the true solution. However, the divergence is observed approximately after 3.5 years from the last observation, which for all practical purpose is sufficient for condition-based maintenance. For stiffness k_6 also, even though the filter estimates are less accurate at earlier time steps, the predicted results manage to give a good estimates of actual value which goes to show that if digital twin is given a regular stream of data, it has the capacity for self correction which in-turn helps better representation of the physical systems.

Bibliography

1. S. Adhikari. Classical normal modes in non-viscously damped linear systems. *AIAA Journal*, 39(5):978–980, May 2001.
2. S. Adhikari. *Structural Dynamic Analysis with Generalized Damping Models: Analysis*. Wiley ISTE, UK, 2013.
3. S. Adhikari and S. Bhattacharya. Dynamic analysis of wind turbine towers on flexible foundations. *Shock and Vibration*, 19(1):37–56, 2012.
4. W.R. Ali and M. Prasad. Piezoelectric mems based acoustic sensors: A review. *Sensors and Actuators A: Physical*, 301:111756, 2020.
5. M. Alvarez and N.D. Lawrence. Sparse convolved gaussian processes for multi-output regression. *Advances in neural information processing systems*, pages 57–64, 2009.
6. M.A. Álvarez and N.D. Lawrence. Computationally efficient convolved multiple output gaussian processes. *Advances in neural information processing systems*, 12:1459–1500, 2011.
7. Arup. Digital twin towards a meaningful framework. *http://www.arup.com/digitaltwinreport*, Tech. rep., Arup, London, England, 2019.
8. Siu-Kui Au and James L Beck. A new adaptive importance sampling scheme for reliability calculations. *Structural safety*, 21(2):135–158, 1999.
9. Siu-Kui Au and James L Beck. Estimation of small failure probabilities in high dimensions by subset simulation. *Probabilistic engineering mechanics*, 16(4):263–277, 2001.
10. Siu-Kui Au and Yu Wang. *Engineering risk assessment with subset simulation*. John Wiley & Sons, 2014.
11. Z. Bai. Krylov subspace techniques for reduced-order modeling of large-scale dynamical systems. *Applied Numerical Mathematics*, 43(1-2):9–44, 2002. cited By (since 1996)170.
12. Z. Bai and Y. Su. Soar: A second-order Arnoldi method for the solution of the quadratic eigenvalue problem. *SIAM Journal on Matrix Analysis and Applications*, 26(3):640–659, 2005.
13. H.T. Banks and D.T. Inman. On damping mechanisms in beams. *Transactions of ASME, Journal of Applied Mechanics*, 58:716–723, 1991.
14. Klaus-Jurgen Bathe. *Finite Element Procedures*. Prentice Hall Inc., Englewood Cliffs, New Jersey, USA, 1995.
15. J. L. Beck and K M. Zuev. Rare event simulation. In *Handbook of Uncertainty Quantification*, pages 1–26. Springer, 2017.
16. Richard Bellman. *Introduction to Matrix Analysis*. McGraw-Hill, New York, USA, 1960.

17. M. Bengisu and Ferrara M. *Materials that move: smart materials, intelligent design*. Springer, 2018.

18. I. Bilionis and N. Zabaras. Multi-output local gaussian process regression: Applications to uncertainty quantification. *Journal of Computational Physics*, 231(17):5718–5746, 2012.

19. I. Bilionis, N. Zabaras, B.A. Konomi, and G. Lin. Multi-output separable Gaussian process: Towards an efficient, fully bayesian paradigm for uncertainty quantification. *Journal of Computational Physics*, 241:212–239, 2013.

20. S. Biswas, S. Chakraborty, S. Chandra, and I. Ghosh. Kriging-based approach for estimation of vehicular speed and passenger car units on an urban arterial. *Journal of Transportation Engineering, Part A: Systems*, 143(3):04016013, 2017.

21. W. Booyse, D.N. Wilke, and S. Heyns. Deep digital twins for detection, diagnostics and prognostics. *Mechanical Systems and Signal Processing*, 140:106612, 2020.

22. J.M. Bourinet, F. Deheeger, and M. Lemaire. Assessing small failure probabilities by combined subset simulation and support vector machines. *Structural Safety*, 33(6):343–353, 2011.

23. G.E. Box and D.A. Pierce. Distribution of residual autocorrelations in autoregressive-integrated moving average time series models. *Journal of the American statistical Association*, 65(332):1509–1526, 1970.

24. Dongxing Cao, Xiangying Guo, and Wenhua Hu. A novel low-frequency broadband piezoelectric energy harvester combined with a negative stiffness vibration isolator. *Journal of Intelligent Material Systems and Structures*, 30(7):1105–1114, 2019.

25. T. K. Caughey. Classical normal modes in damped linear dynamic systems. *Transactions of ASME, Journal of Applied Mechanics*, 27:269–271, June 1960.

26. T. K. Caughey and M. E. J. O'Kelly. Classical normal modes in damped linear dynamic systems. *Transactions of ASME, Journal of Applied Mechanics*, 32:583–588, September 1965.

27. S. Chakraborty. Simulation free reliability analysis: A physics-informed deep learning based approach. *arXiv preprint arXiv:2005.01302*, 2020.

28. S. Chakraborty. Transfer learning based multi-fidelity physics informed deep neural network. *Journal of Computational Physics*, 426:109942, 2021.

29. S. Chakraborty and S. Adhikari. Machine learning based digital twin for dynamical systems with multiple time-scales. *Computers & Structures*, 243:106410, 2021.

30. S. Chakraborty, S. Adhikari, and R. Ganguli. The role of surrogate models in the development of digital twins of dynamic systems. *Applied Mathematical Modelling*, 90:662–681, 2021.

31. S. Chakraborty, T. Chatterjee, R. Chowdhury, and S. Adhikari. A surrogate based multi-fidelity approach for robust design optimization. *Applied Mathematical Modelling*, 47, 2017.

32. S Chakraborty and R Chowdhury. Polynomial Correlated Function Expansion for Nonlinear Stochastic Dynamic Analysis. *Journal of Engineering Mechanics*, 141(3):04014132:1—-04014132:11, 2014.

33. S Chakraborty and R Chowdhury. A semi-analytical framework for structural reliability analysis. *Computer Methods in Applied Mechanics and Engineering*, 289(1):475–497, Feb 2015.

34. S Chakraborty and R Chowdhury. Multivariate Function Approximations Using D-MORPH Algorithm. *Applied Mathematical Modelling*, 39(23-24):7155–7180, 2015.

35. S Chakraborty and R Chowdhury. Assessment of polynomial correlated function expansion for high-fidelity structural reliability analysis. *Structural Safety*, 59:9–19, 2016.

36. S Chakraborty and R Chowdhury. An efficient algorithm for building locally refined hp – adaptive H-PCFE: Application to uncertainty quantification. *Journal of Computational Physics*, 351:59–79, 2017.

37. S. Chakraborty and R. Chowdhury. Hybrid framework for the estimation of rare failure event probability. *Journal of Engineering Mechanics*, 143(5):04017010, 2017.

38. S Chakraborty and R Chowdhury. Polynomial correlated function expansion. In *Modeling and simulation techniques in structural engineering*, pages 348–373. IGI global, 2017.

39. S. Chakraborty and R. Chowdhury. Polynomial correlated function expansion. *Modeling and simulation techniques in structural engineering, IGI global*, pages 348–373, 2017.

40. S. Chakraborty and R. Chowdhury. Graph-theoretic-approach-assisted gaussian process for nonlinear stochastic dynamic analysis under generalized loading. *Journal of Engineering Mechanics*, 145(12):04019105, 2019.

41. S Chakraborty, B Mandal, R Chowdhury, and A Chakrabarti. Stochastic free vibration analysis of laminated composite plates using polynomial correlated function expansion. *Composite Structures*, 135:236–249, 2016.

42. S. Chakraborty and N. Zabaras. Efficient data-driven reduced-order models for high-dimensional multiscale dynamical systems. *Computer Physics Communications*, 230:70–88, 2018.

43. Subrata Chakraborty and Bijan Kumar Roy. Reliability based optimum design of Tuned Mass Damper in seismic vibration control of structures with bounded uncertain parameters. *Probabilistic Engineering Mechanics*, 26(2):215–221, 2011.

44. T Chatterjee, S Chakraborty, and R Chowdhury. Analytical moment based approximation for robust design optimization. *Structural and Multidisciplinary Optimization*, 58(5):2135–2162, 2018.

45. M. Chen, S. Mao, , and Y. Liu. Big data: A survey. *Mobile networks and application*, 19(2):171–209, 2014.

46. Min Chen, Xiaobo Shi, Yin Zhang, Di Wu, and Mohsen Guizani. Deep features learning for medical image analysis with convolutional autoencoder neural network. *IEEE Transactions on Big Data*, 2017.

47. R Chowdhury, B N Rao, and A M Prasad. High dimensional model representation for piece-wise continuous function approximation. *Communications in Numerical Methods in Engineering*, 24(12):1587–1609, 2007.

48. R Chowdhury, B N Rao, and A M Prasad. High-dimensional model representation for structural reliability analysis. *Communications in Numerical Methods in Engineering*, 25(4):301–337, apr 2009.

49. D. Chronopoulos, M. Ichchou, B. Troclet, and O. Bareille. Predicting the broadband vibroacoustic response of systems subject to aeroacoustic loads by a krylov subspace reduction. *Applied Acoustics*, 74(12):1394 – 1405, 2013.

50. C. Cimino, E. Negri, and L. Fumagalli. Review of digital twin applications in manufacturing. *Computers in Industry*, 113:103130, 2019.

51. C Allin Cornell. Bounds on the reliability of structural systems. *Journal of the Structural Division*, 93(1):171–200, 1967.

52. P.D.U. Coronado, R. Lynn, W. Louhichi, M. Parto, T. Wescoat, and T. Kurfess. Part data integration in the shop floor digital twin: Mobile and cloud technologies to enable a manufacturing execution system. *Journal of Manufacturing Systems*, 48:25–33, 2018.

53. C. Cremona and J. Santos. Structural health monitoring as a big-data problem. *Structural Engineering International*, 28(3):243–254, 2018.

54. G Cybenko. Approximation by superpositions of a sigmoidal function. *Mathematics of Control, Signals and Systems*, 5(4):455–455, 1992.

55. M. D. Buhmann. Radial basis functions. *Acta Numerica 2000*, 9:1–38, 2000.

56. H. Dai, H. Zhang, and W. Wang. A multiwavelet neural network-based response surface method for structural reliability analysis. *Computer-Aided Civil and Infrastructure Engineering*, 30(2):151–162, 2015.

57. S Das, S Chakraborty, Y Chen, and S Tesfamariam. Robust design optimization for sma based nonlinear energy sink with negative stiffness and friction. *Soil Dynamics and Earthquake Engineering*, 140:106466, 2021.

58. T. Debroy, W. Zhang, J. Turner, and S. Babu. Building digital twins of 3d printing machines. *Scripta Materialia*, 135:119–124, 2017.

59. F.A. DiazDelaO and S. Adhikari. Structural dynamic analysis using gaussian process emulators. *Engineering Computations*, 27(5):580–605, 2010.

60. F.A. DiazDelaO and S. Adhikari. Gaussian process emulators for the stochastic finite element method. *International Journal of Numerical Methods in Engineering*, 87(6):521–540, 2011.

61. F.A. DiazDelaO and S. Adhikari. Bayesian assimilation of multi-fidelity finite element models. *Computers and Structures*, 92-93(2):206–215, 2012.

62. O Ditlevsen, R E Melchers, and H Gluver. General multi-dimensional probability integration by directional simulation. *Computers & Structures*, 36(2):355–368, 1990.

63. A. Doucet, S. Godsill, and C. Andrieu. On sequential monte carlo sampling methods for bayesian filtering. *Statistics and computing*, 10(3):197–208, 2000.

64. X Du and W Chen. Sequential optimization and reliability assessment method for efficient probabilistic design. *Journal of Mechanical Design*, 126(2):225–233, 2004.

65. V. Dubourg. *Adaptive surrogate models for reliability analysis and reliability-based design optimization*. Ph.D. thesis, Universite Blaise Pascal, Clermont-Ferrand, France, 2011.

66. I. Enevoldsen and J.D. Sørensen. Reliability-based optimization in structural engineering. *Structural Safety*, 15(3):169–196, 1994.

67. S Engelund and R Rackwitz. A benchmark study on importance sampling techniques in structural reliability. *Structural safety*, 12(4):255–276, 1993.

68. I. Errandonea, S. Beltrán, and S. Arrizabalaga. Digital twin for maintenance: A literature review. *Computers in Industry*, 123:103316, 2020.

69. G. Evensen. *Data Assimilation: The Ensemble Kalman Filter*. Springer, Berlin, 2006.

70. D.J. Ewins. *Modal Testing: Theory and Practice*. Research Studies Press, Taunton, 1984.

71. D. Feng, M. Feng, E. Ozer, and Y. Fakuda. A vision-based sensor for noncontact structural displacement measurement. *Sensors*, 15(7):16557–16575, 2015.

72. K Fukushima. A self-organizing neural network model for a mechanism of pattern recognition unaffected by shift in position. *Biol. Cybern*, 36:193–202, 1980.

73. R. Ganguli. Fuzzy logic intelligent system for gas turbine module and system fault isolation. *Journal of propulsion and power*, 18(2):440–447, 2002.

74. R. Ganguli. Noise and outlier removal from jet engine health signals using weighted fir median hybrid filters. *Mechanical Systems and Signal Processing*, 16(6):967–978, 2002.

75. R. Ganguli. Jet engine gas-path measurement filtering using center weighted idempotent median filters. *Journal of propulsion and power*, 19(5):930–937, 2003.

76. R Ganguli. *Gas turbine diagnostics*. CRC Press, Boca Raton, FL, USA, 2013.

77. R. Ganguli and S. Adhikari. The digital twin of discrete dynamic systems: Initial approaches and future challenges. *Applied Mathematical Modelling*, 77:1110–1128, 2020.

78. R. Ganguli, I. Chopra, and D.J. Haas. Simulation of helicopter rotorsystem structural damage, blade mistracking, friction, and freeplay. *Journal of aircraft*, 35(4):591–597, 1998.

79. S. Garg, A. Gogoi, S. Chakraborty, and B. Hazra. Machine learning based digital twin for stochastic nonlinear multi-degree of freedom dynamical system. *Probabalistic Engineering Mechanics*, 66:103173, 2021.

80. N Geneva and N Zabaras. Modeling the dynamics of pde systems with physics-constrained deep auto-regressive networks. *Journal of Computational Physics*, 403:109056, 2020.

81. M. Géradin and D. Rixen. *Mechanical Vibrations*. John Wiely & Sons, New York, NY, second edition, 1997. Translation of: Théorie des Vibrations.

82. R. Ghanem and P. Spanos. *Stochastic Finite Elements: A Spectral Approach*. Springer-Verlag, New York, USA, 1991.

83. G. Gillich and I. Mituletu. Signal post-processing for accurate evaluation of the natural frequencies. *Structural Health Monitoring, Springer*, pages 13–37, 2017.

84. K. Glaessgen and D. Stargel. Multiphysics stimulated simulation digital twin methods for fleet management. *53rd AIAA/ASME/ASCE/AHS/ASC Structures, Structural Dynamics and Materials Conference*, 1818, 2012.

85. S. Goswami, C. Anitescu, S. Chakraborty, and T. Rabczuk. Transfer learning enhanced physics informed neural network for phase-field modeling of fracture. *Theoretical and Applied Fracture Mechanics*, 106:102447, 2020.

86. S. Goswami, S. Chakraborty, R. Chowdhury, and T. Rabczuk. Threshold shift method for reliability-based design optimization. *Structural and Multidisciplinary Optimization*, 60(5):2053–2072, 2019.

87. S.R. Gunn. *Technical Report: Support vector machines for classification and regression*. University of Southampton, Southampton, U.K., 1997.

88. H. H. Millwater, J. Ocampo, and N. Crosby. Probabilistic methods for risk assessment of airframe digital twin structures. *Engineering Fracture Mechanics*, 221:106674, 2019.

89. S. Haag and R. Anderl. Digital twin – proof of concept. *Manufacturing Letters*, 15:64–66, 2018.

90. T.W. Hall. Artificial gravity in theory and practice. *46th International Conference on Environmental Systems*, 2016.

91. B. He and K.J. Bai. Digital twin-based sustainable intelligent manufacturing: a review. *Advances in Manufacturing*, 9(1):1–21, 2021.

92. S. Hoodorozhkov and A. Krasilnikov. Digital twin of wheel tractor with automatic gearbox. *E3S Web of Conferences, EDP Sciences*, 164:p. 03032, 2020.

93. G Hooker. Generalized Functional ANOVA Diagnostics for High-Dimensional Functions of Dependent Variables. *Journal of Computational and Graphical Statistics*, 16(3):709–732, sep 2007.

94. B Huang and X Du. Analytical robustness assessment for robust design. *Structural and Multidisciplinary Optimization*, 34(2):123–137, 2007.

95. S Huang, G Wang, Y Yan, and X Fang. Blockchain based data management for digital twin of product. *Journal of Manufacturing Systems*, 54:361–371, 2020.

96. E Hüllermeier and W Waegeman. Aleatoric and epistemic uncertainty in machine learning: An introduction to concepts and methods. *Machine Learning*, pages 1–50, 2021.

97. Inman D J. *Engineering Vibration*. Prentice Hall PTR, NJ, USA, 2003.

98. M. Javaid, A. Haleem, S. Rab, R.P. Singh, and R. Suman. Sensors for daily life: a review. *Sensors International*, 2:100121, 2021.

99. M. Kamiński. Stochastic perturbation approach to the wavelet-based analysis. *Numerical Linear Algebra with Applications*, 11(4):355–370, 2004.

100. M G Kapteyn, D J Knezevic, and K Willcox. Toward predictive digital twins via component-based reduced-order models and interpretable machine learning. In *AIAA Scitech 2020 Forum*, page 0418, 2020.

101. M.G. Kapteyn, K.E. Willcox, and A.B. Philpott. Distributionally robust optimization for engineering design under uncertainty. *International Journal for Numerical Methods in Engineering*, 120(7):835–859, 2019.

102. S Karumuri, R Tripathy, I Bilionis, and J Panchal. Simulator-free solution of high-dimensional stochastic elliptic partial differential equations using deep neural networks. *Journal of Computational Physics*, 404:109120, 2020.

103. M.J. Kaur, V.P. Mishra, and P. Maheshwari. The convergence of digital twin, iot, and machine learning: Transforming data into action. *Digital Twin Technologies and Smart Cities, Springer*, pages 3–17, 2020.

104. M. Khalil, A. Sarkar, and S. Adhikari. Nonlinear filters for chaotic oscillatory systems. *Nonlinear Dynamics*, 55(1-2):113–137, 2009.

105. P Kim. *Kalman Filters for Beginners*. Createspace Independent Publishing Platform, 2011.

106. P.S. Koutsourelakis and E. Bilionis. Scalable bayesian reduced-order models for simulating high-dimensional multiscale dynamical systems. *Multiscale Modeling & Simulation*, 9(1):449–485, 2011.

107. E Kreyszig. *Advanced engineering mathematics*. John Wiley & Sons, New York, eigth edition, 1999.

108. Y Kumar, P Bahl, and S Chakraborty. State estimation with limited sensors–a deep learning based approach. *Journal of Computational Physics*, 457:111081, 2022.

109. A. Kundu and S. Adhikari. Dynamic analysis of stochastic structural systems using frequency adaptive spectral functions. *Probabilistic Engineering Mechanics*, 39(1):23–38, 2015.

110. S Lawrence, C L Giles, A C Tsoi, and A D Back. Face recognition: A convolutional neural-network approach. *IEEE transactions on neural networks*, 8(1):98–113, 1997.

111. D Lazzaro and L B Montefusco. Radial basis functions for the multivariate interpolation of large scattered data sets. *Journal of Computational and Applied Mathematics*, 140(1-2):521–536, 2002.

112. B Li, T Bengtsson, and P Bickel. Curse-of-dimensionality revisited: Collapse of importance sampling in very large scale systems. *Rapport technique*, 85:205, 2005.

113. G Li, J Hu, S-W Wang, P G Georgopoulos, J Schoendorf, and H Rabitz. Random sampling-high dimensional model representation (RS-HDMR) and orthogonality of its different order component functions. *The journal of physical chemistry. A*, 110(7):2474–2485, 2006.

114. G Li and H Rabitz. D-MORPH regression: application to modeling with unknown parameters more than observation data. *Journal of Mathematical Chemistry*, 48(4):1010–1035, 2010.

115. G Li and H Rabitz. D-MORPH regression: application to modeling with unknown parameters more than observation data. *Journal of Mathematical Chemistry*, 48(4):1010–1035, 2010.

116. G Li and H Rabitz. General formulation of HDMR component functions with independent and correlated variables. *Journal of Mathematical Chemistry*, 50(1):99–130, 2012.

117. G Li and H Rabitz. General formulation of HDMR component functions with independent and correlated variables. *Journal of Mathematical Chemistry*, 50(1):99–130, 2012.

118. G Li, H Rabitz, J Hu, Z Chen, and Y Ju. Regularized random-sampling high dimensional model representation (RS-HDMR). *Journal of Mathematical Chemistry*, 43(3):1207–1232, sep 2007.

119. G Li, R Rey-de Castro, and H Rabitz. D-MORPH regression for modeling with fewer unknown parameters than observation data. *Journal of Mathematical Chemistry*, 50(7):1747–1764, 2012.

120. J Liang, Z P. Mourelatos, and J Tu. A single-loop method for reliability-based design optimisation. *International Journal of Product Development*, 5(1/2):76, 2008.

121. J.S. Liu and R. Chen. Blind deconvolution via sequential imputations. *Journal of the American Statistical Association*, 90(430):567–576, 1995.

122. L. Liu, W. Hua, and Y. Lei. Real-time simultaneous identification of structural systems and unknown inputs without collocated acceleration measurements based on mekf-ui. *Measurement*, 122:545–553, 2018.

123. Y. Lu, C. Liu, I. Kevin, K. Wang, H. Huang, and X. Xu. Digital twin-driven smart manufacturing: Connotation, reference model, applications and research issues. *Robotics and Computer-Integrated Manufacturing*, 61(Art. 101837):64–66, 2020.

124. M. M. Zhou, J. J. Yan, and D. Feng. Digital twin framework and its application to power grid online analysis. *CSEE Journal of Power and Energy Systems*, 5(3):391–398, 2019.

125. H. Mao and S. Mahadevan. Fatigue damage modelling of composite materials. *Composite Structures*, 58(4):405–410, 2002.

126. S D Marchi and G Santin. A new stable basis for radial basis function interpolation. *Journal of Computational and Applied Mathematics*, 253:1–13, 2013.

127. C.G. Mattera, J. Quevedo, T. Escobet, H.R. Shaker, and M. Jradi. A method for fault detection and diagnostics in ventilation units using virtual sensors. *Sensors*, 18(11):3931, 2018.

128. L. Meirovitch. *Analytical Methods in Vibrations*. Macmillan Publishing Co., Inc., New York, 1967.

129. L. Meirovitch. *Computational Methods in Structural Dynamics*. Sijthoff & Noordohoff, Netherlands, 1980.

130. L. Meirovitch. *Principles and Techniques of Vibrations*. Prentice-Hall International, Inc., New Jersey, 1997.

131. S K Mishra, B K Roy, and S Chakraborty. Reliability-based-design-optimization of base isolated buildings considering stochastic system parameters subjected to random earthquakes. *International Journal of Mechanical Sciences*, 75:123–133, 2013.

132. P.D. Moral, A. Doucet, and A. Jasra. Sequential monte carlo samplers. *Journal of the Royal Statistical Society: Series B (Statistical Methodology)*, 68(3):411–436, 2006.

133. T. Mukhopadhyay, S. Chakraborty, S. Dey, S. Adhikari, and R. Chowdhury. A critical assessment of kriging model variants for high-fidelity uncertainty quantification in dynamics of composite shells. *Archives of Computational Methods in Engineering*, 24(3):495–518, 2017.

134. K.P. Murphy. *Machine learning: a probabilistic perspective*. MIT Press, Cambridge, MA, 2012.

135. R. Nayek, S. Chakraborty, and S. Narasimhan. A gaussian process latent force model for joint input-state estimation in linear structural systems. *Mechanical Systems and Signal Processing*, 128:497–530, 2019.

136. A.H. Nayfeh. *Introduction to Perturbation Techniques*. John Wiley & Sons, New York, NY, 1993.

137. A.H. Nayfeh and D.T. Mook. *Nonlinear oscillations*. John Weily & Sons, New York, NY, 1979.

138. D. E. Newland. *Mechanical Vibration Analysis and Computation*. Longman, Harlow and John Wiley, New York, 1989.

139. N.C. Nigam. *Introduction to Random Vibration*. The MIT Press, Cambridge, Massachusetts, 1983.

140. S.J. Orfanidis. *Introduction to Signal Processing*. Prentice Hall, 2010.

141. E. Oztemel and S. Gursev. Literature review of industry 4.0 and related technologies. *Journal of Intelligent Manufacturing*, 31(1):127–182, 2020.

142. A. Papoulis and S.U. Pillai. *Probability, Random Variables and Stochastic Processes*. McGraw-Hill, Boston, USA, 2002.

143. K.T. Park, D. Lee, and S.D. Noh. Operation procedures of a work-center-level digital twin for sustainable and smart manufacturing. *International*

Journal of Precision Engineering and Manufacturing-Green Technology, 7(3):791–814, 2020.

144. M. Paz. *Structural Dynamics: Theory and Computation*. Van Nostrand, Reinhold, second edition, 1980.

145. M. Pelikan, D.E. Goldberg, and E. Cantú-Paz. The bayesian optimization algorithm. *Proceedings of the 1st Annual Conference on Genetic and Evolutionary Computation-Volume 1, Morgan Kaufmann Publishers Inc.*, pages 525–532, 1999.

146. P. Perdikaris, D. Venturi, J.O. Royset, and G.E. Karniadakis. Multi-fidelity modelling via recursive co-kriging and gaussian–markov random fields. *Proceedings of the Royal Society A: Mathematical, Physical and Engineering Sciences*, 471(2179):20150018, 2015.

147. S.E. Pryse, S. Adhikari, and A. Kundu. Sample-based and sample-aggregated galerkin projection schemes for structural dynamics. *Probabilistic Engineering Mechanics*, 54(10):118–130, 2018.

148. S. Rahman and D. Wei. Reliability-based design optimization by a univariate decomposition method. *13th AIAA/ISSMO Multidisciplinary Analysis Optimization Conference*, AIAA 2010-9037:1–14, 2010.

149. M Raissi and G E Karniadakis. Hidden physics models: Machine learning of nonlinear partial differential equations. *Journal of Computational Physics*, 357:125–141, 2018.

150. M Raissi, P Perdikaris, and G E Karniadakis. Physics-informed neural networks: A deep learning framework for solving forward and inverse problems involving nonlinear partial differential equations. *Journal of Computational Physics*, 378:686–707, 2019.

151. A. Rakhlin. Convolutional neural networks for sentence classification. *Github*, 2016.

152. C R Rao and S K Mitra. *Generalized inverse of matrix and its applications*. Wiley, New York, U.S.A., 1971.

153. S. S. Rao. *Mechanical Vibrations*. Prentice Hall PTR, NJ, USA, 2011.

154. A. Rasheed, O. San, and T. Kvamsdal. Digital twin: Values, challenges and enablers from a modeling perspective. *IEEE Access*, 8:21980–22012, 2020.

155. C E Rasmussen. Gaussian processes in machine learning. In *Summer school on machine learning*, pages 63–71. Springer, 2003.

156. C.E. Rasmussen and H. Nickisch. Gaussian processes for machine learning (gpml) toolbox. *Journal of machine learning research*, 11:3011–3015, 2010.

157. J. W. Rayleigh. *Theory of Sound (two volumes)*. Dover Publications, New York, 1945 re-issue, second edition, 1877.

158. K. Reifsnider and P. Majumdar. Multiphysics stimulated simulation digital twin methods for fleet management. *54th AIAA/ASME/ASCE/AHS/ASC Structures, Structural Dynamics, and Materials Conference*, 1578, 2013.

159. X Ren and S Rahman. Robust design optimization by adaptive-sparse polynomial dimensional decomposition. In *Volume 1B: 36th Computers and Information in Engineering Conference*. American Society of Mechanical Engineers, 2016.

160. J.B. Roberts and P.D. Spanos. *Random Vibration and Statistical Linearization*. John Wiley and Sons Ltd, Chichester, England, 1990, 1990.

161. Debasish Roy and G Visweswara Rao. *Stochastic dynamics, filtering and optimization*. Cambridge University Press, 2017.

162. N. Roy and R. Ganguli. Helicopter rotor blade frequency evolution with damage growth and signal processing. *Journal of Sound and Vibration*, 283(3-5):821–851, 2005.

163. Reuven Y Rubinstein and Dirk P Kroese. *Simulation and the Monte Carlo method*, volume 10. John Wiley & Sons, 2016.

164. T. Ruohomäki, E. Airaksinen, P. Huuska, O. Kesäniemi, M. Martikka, and J. Suomisto. Smart city platform enabling digital twin. *IEEE International Conference on Intelligent Systems (IS)*, pages 155–161, 2018.

165. A. Sabato, C. Niezrecki, and G. Fortino. Wireless mems-based accelerometer sensor boards for structural vibration monitoring: a review. *IEEE Sensors Journal*, 17(2):2226–235, 2016.

166. A. Saha, S. Chakraborty, S. Chandra, and I. Ghosh. Kriging based saturation flow models for traffic conditions in indian cities. *Journal of Transportation Engineering, Part A: Systems*, 118:38–51, 2018.

167. S.E. Said and D.A. Dickey. Testing for unit roots in autoregressive-moving average models of unknown order. *Biometrika*, 71(3):599–607, 1984.

168. E. Samaniego, C. Anitescu, S. Goswami, V.M. Nguyen-Thanh, H. Guo, X. Hamdia, K. Zhuang, and T. Rabczuk. An energy approach to the solution of partial differential equations in computational mechanics via machine learning: Concepts, implementation and applications. *Computer Methods in Applied Mechanics and Engineering*, 362:112790, 2020.

169. S. Sarkar, K.K. Reddy, M. Giering, and M.R. Gurvich. Deep learning for structural health monitoring: A damage characterization application. *Annual conference of the prognostics and health management society*, pages 176–182, 2016.

170. S Särkkä. *Bayesian filtering and smoothing*, volume 3. Cambridge University Press, 2013.

171. R Schobi, B Sudret, and J Wiart. Polynomial-chaos-based Kriging. *International Journal for Uncertainty Quantification*, 5(2), 2015.

172. G. Shao and M. Helu. Framework for a digital twin in manufacturing: Scope and requirements. *Manufacturing Letters*, 24:105–107, 2020.

173. L Shi and S P Lin. A new RBDO method using adaptive response surface and first-order score function for crashworthiness design. *Reliability Engineering & System Safety*, 156:125–133, 2016.

174. D Simon. *Optimal State Estimation*. Wiley Interscience, 2006.

175. A.P. Singh, V. Mani, and R. Ganguli. Genetic programming metamodel for rotating beams. *Computer Modeling in Engineering and Sciences*, 21(2):133, 2007.

176. V. Singh and K.E. Willcox. Engineering design with digital thread. *AIAA Journal*, 56(11):4515–4528, 2018.

177. E. Snelson and Z. Ghahramani. Sparse gaussian processes using pseudo-inputs. *Advances in neural information processing systems*, pages 1257–1264, 2006.

178. I M Sobol. Sensitivity estimates for nonlinear mathematical models. *Mathematical Modeling and Computational Experiment*, 1(4):407–414, 1993.

179. S. Sony, S. Laventure, and A. Sadhu. A literature review of next-generation smart sensing technology in structural health monitoring. *Structural Control and Health Monitoring*, 26(3):e2321, 2019.

180. V Souza, R Cruz, W Silva, S Lins, and V Lucena. A digital twin architecture based on the industrial internet of things technologies. In *2019 IEEE International Conference on Consumer Electronics (ICCE)*, pages 1–2. IEEE, 2019.

181. B. Sudret. Global sensitivity analysis using polynomial chaos expansions. *Reliability Engineering & System Safety*, 93(7):964–979, 2008.

182. F. Tao, F. Sui, A. Liu, Q. Qi, M. Zhang, B. Song, S. Guo, C.Y. Lu, and A. Nee. Digital twin-driven product design framework. *International Journal of Production Research*, 57(12):3935–3953, 2019.

183. F. Tao, H. Zhang, A. Liu, and A. Nee. Digital twin in industry: state of the art. *IEEE Transactions on Industrial Informatics*, 15(4):2405–2415, 2018.

184. R Thakur and KB Misra. Monte Carlo simulation for reliability evaluation of complex systems. *International Journal of Systems Science*, 9(11):1303–1308, 1978.

185. Tapas Tripura, Ankush Gogoi, and Budhaditya Hazra. An ito–taylor weak 3.0 method for stochastic dynamics of nonlinear systems. *Applied Mathematical Modelling*, 86:115–141, 2020.

186. J. Tu, K. K. Choi, and Y. H. Park. A New Study on Reliability-Based Design Optimization. *Journal of Mechanical Design*, 121(4):557–564, 1999.

187. E. Tuegel, A. Ingraffea, T. Eason, and S. Spottswood. Reengineering aircraft structural life prediction using a digital twin. *International Journal of Aerospace Engineering*, Art. 154768, 2011.

188. J. Vetelino and A. Reghu. *Introduction to Sensors*. CRC Press, Boca Raton, FL, USA, 2017.

189. E A Wan and R Van Der Merwe. The unscented kalman filter for nonlinear estimation. In *Proceedings of the IEEE 2000 Adaptive Systems for Signal Processing, Communications, and Control Symposium (Cat. No. 00EX373)*, pages 153–158. Ieee, 2000.

190. J. Wan and N. Zabaras. A probabilistic graphical model approach to stochastic multiscale partial differential equations. *Journal of Computational Physics*, 250:477–510, 2013.

191. J. Wang, L. Ye, R. Gao, C. Li, and L. Zhang. Digital twin for rotating machinery fault diagnosis in smart manufacturing. *International Journal of Production Research*, 57(12):3920–3934, 2019.

192. S W Wang, P G Georgopoulos, G Li, and H Rabitz. Random sampling-high dimensional model representation (RS-HDMR) with nonuniformly distributed variables: application to an integrated multimedia multi-pathway exposure and dose model for trichloroethylene. *The Journal of Physical Chemistry A*, 107(23):4707–4718, 2003.

193. W. Wang, J. Huang, Q. Liu, and Z. Zhou. Dynamic strain measurement of hydraulic system pipeline using fibre bragg grating sensors. *Advances in Mechanical Engineering*, 8(4), 2016.

194. Website. https://aws.amazon.com/emr/details/hadoop/what-is-hadoop.

195. Website. https://aws.amazon.com/well-architected-tool.

196. Website. https://datatracker.ietf.org/doc/html/rfc5673.

197. Website. https://docs.microsoft.com/en-us/azure/digital-twins/concepts-models.

198. Website. https://github.com/azure-samples/digital-twins-docs-code.

199. Website. https://ieeexplore.ieee.org/abstract/document/8258937.

200. Website. https://ieeexplore.ieee.org/document/9711981.

201. Website. https://www.cio.com/article/193822/digital-twins-and-the-enterprise-edge.html.

202. Website. https://www.ibm.com/blogs/aws/building-the-digital-representation-with-digital-twin-using-aws-stack.

203. Website. https://www.leanix.net/en/blog/iot-devices-sensors-and-actuators-explained.

204. Website. https://www.youtube.com/watch?v=vdg7ukq2ekk&t=199s.

205. G Welch and G Bishop. An introduction to the kalman filter, 1995.

206. N Wiener. The homogeneous chaos. *American Journal of Mathematics*, 60(4):897–936, 1938.

207. J. H. Wilkinson. *The Algebraic Eigenvalue Problem*. Oxford University Press, Oxford, UK, 1988.

208. C.K. Williams and C.E. Rasmussen. *Gaussian processes for machine learning*. MIT Press, Cambridge, MA, 2006.

209. A Wirgin. The inverse crime. *arXiv preprint math-ph/0401050*, 2004.

210. K. Worden, E. Cross, P. Gardner, R. Barthorpe, and D. Wagg. On digital twins, mirrors and virtualisations. *Model Validation and Uncertainty Quantification*, 3(1):285–295, 2020.

211. F. Wu and W.X. Zhong. A modified stochastic perturbation method for stochastic hyperbolic heat conduction problems. *Computer Methods in Applied Mechanics and Engineering*, 305:739–758, 2016.

212. S Xiao, S Oladyshkin, and W Nowak. Reliability analysis with stratified importance sampling based on adaptive kriging. *Reliability Engineering & System Safety*, 197:106852, 2020.

213. M Xiong, F C Arnett, X Guo, H Xiong, and X Zhou. Differential dynamic properties of scleroderma fibroblasts in response to perturbation of environmental stimuli. *PloS One*, 3(2):e1693:1–e1693:12, 2008.

214. D. Xiu and G.E. Karniadakis. The wiener-askey polynomial chaos for stochastic differential equations. *SIAM Journal on Scientific Computing*, 24(2):619–644, 2002.

215. Dongbin Xiu and George Em Karniadakis. The wiener–askey polynomial chaos for stochastic differential equations. *SIAM journal on scientific computing*, 24(2):619–644, 2002.

216. H. Xu, H. Chang, and D. Zhang. Deep-learning based data-driven discovery of partial differential equations from discrete and noisy data. *Communications in Computational Physics*, 29:698–728, 2021.

217. Y. Xu and A. Helal. Scalable cloud-sensor architecture for the internet of things. *IEEE Internet of Things Journal*, 3(3):285–298, 2015.

218. Y. Xu, A. C. Helal, C. Li, S. Mahadevan, Y. Ling, S. Choze, and L. Wang. Dynamic bayesian network for aircraft health monitoring digital twin. *AIAA Journal*, 55(3):930–941, 2017.

219. B.D. Youn, K.K. Choi, R.-J. Yang, and L. Gu. Reliability-based design optimization for crashworthiness of vehicle side impact. *Structural and Multidisciplinary Optimization*, 26(3-4):272–283, 2004.

220. N Young. *An introduction to Hilbert space.* Cambridge university press, 1988.

221. J. Yu, B. Yan, X. Meng, X. Shao, and H. Ye. Measurement of bridge dynamic responses using network-based real-time kinematic gnss technique. *Journal of Surveying Engineering*, 142(3):04015013, 2016.

222. P. Zarchan and H. Musoff. *Fundamentals of Kalman Filtering.* AIAA, Reston, Virginia, USA, 2015.

223. W. Zhang, G. Peng, C. Li, Y. Chen, and Z. Zhang. A new deep learning model for fault diagnosis with good anti-noise and domain adaptation ability on raw vibration signals. *Sensors*, 17(2):425, 2017.

224. W. Zhao, T. Tao, E. Zio, and W. Wang. A novel hybrid method of parameters tuning in support vector regression for reliability prediction: particle swarm optimization combined with analytical selection. *IEEE Transactions on Reliability*, pages 1–13, 2016.

225. C. Zhuang, J. Liu, H. Xiong, X. Ding, S. Liu, and G. Weng. Connotation, architecture and trends of product digital twin. *Computer Integrated Manufacturing Systems*, 23(4):753–768, 2017.

226. C. Zhuang, J. Liu, H. Xiong, X. Ding, S. Liu, and G. Weng. Digital twin-driven product design, manufacturing and service with big data. *International Journal of Advanced Manufacturing Technology*, 94(9-12):3563–3576, 2018.

227. K M. Zuev. Subset simulation method for rare event estimation: an introduction. In *Encyclopedia of Earthquake Engineering*, pages 1–25. Springer Berlin Heidelberg, Berlin, Heidelberg, 2013.

Index